T0184383

Applied Mathematical Sciences

EDITORS

Fritz John
Courant Institute of Mathematical Sciences
New York University
New York, NY 10012

J.E. Marsden
Department of Mathematics
University of California
Berkeley, CA 94720

Lawrence Sirovich
Division of Applied Mathematics
Brown University
Providence, RI 02912

ADVISORS

H. Cabannes University of Paris-VI

M. Ghil New York University

J.K. Hale Brown University

J. Keller Stanford University

G.B. Whitham California Inst. of Technology

EDITORIAL STATEMENT

The mathematization of all sciences, the fading of traditional scientific boundaries, the impact of computer technology, the growing importance of mathematical-computer modelling and the necessity of scientific planning all create the need both in education and research for books that are introductory to and abreast of these developments.

The purpose of this series is to provide such books, suitable for the user of mathematics, the mathematician interested in applications, and the student scientist. In particular, this series will provide an outlet for material less formally presented and more anticipatory of needs than finished texts or monographs, yet of immediate interest because of the novelty of its treatment of an application or of mathematics being applied or lying close to applications.

The aim of the series is, through rapid publication in an attractive but inexpensive format, to make material of current interest widely accessible. This implies the absence of excessive generality and abstraction, and unrealistic idealization, but with quality of exposition as a goal.

Many of the books will originate out of and will stimulate the development of new undergraduate and graduate courses in the applications of mathematics. Some of the books will present introductions to new areas of research, new applications and act as signposts for new directions in the mathematical sciences. This series will often serve as an intermediate stage of the publication of material which, through exposure here, will be further developed and refined. These will appear in conventional format and in hard cover.

MANUSCRIPTS

The Editors welcome all inquiries regarding the submission of manuscripts for the series. Final preparation of all manuscripts will take place in the editorial offices of the series in the Division of Applied Mathematics, Brown University, Providence, Rhode Island.

SPRINGER-VERLAG NEW YORK INC., 175 Fifth Avenue, New York, N. Y. 10010

Printed in U.S.A.

Applied Mathematical Sciences | Volume 56

Applied Mathematical Sciences

1. John: **Partial Differential Equations**, 4th ed.
2. Sirovich: **Techniques of Asymptotic Analysis.**
3. Hale: **Theory of Functional Differential Equations**, 2nd ed.
4. Percus: **Combinatorial Methods.**
5. von Mises/Friedrichs: **Fluid Dynamics.**
6. Freiberger/Grenander: **A Short Course in Computational Probability and Statistics.**
7. Pipkin: **Lectures on Viscoelasticity Theory.**
8. Giacaglia: **Perturbation Methods in Non-Linear Systems.**
9. Friedrichs: **Spectral Theory of Operators in Hilbert Space.**
10. Stroud: **Numerical Quadrature and Solution of Ordinary Differential Equations.**
11. Wolovich: **Linear Multivariable Systems.**
12. Berkovitz: **Optimal Control Theory.**
13. Bluman/Cole: **Similarity Methods for Differential Equations.**
14. Yoshizawa: **Stability Theory and the Existence of Periodic Solutions and Almost Periodic Solutions.**
15. Braun: **Differential Equations and Their Applications**, 3rd ed.
16. Lefschetz: **Applications of Algebraic Topology.**
17. Collatz/Wetterling: **Optimization Problems.**
18. Grenander: **Pattern Synthesis: Lectures in Pattern Theory, Vol I.**
19. Marsden/McCracken: **The Hopf Bifurcation and its Applications.**
20. Driver: **Ordinary and Delay Differential Equations.**
21. Courant/Friedrichs: **Supersonic Flow and Shock Waves.**
22. Rouche/Habets/Laloy: **Stability Theory by Liapunov's Direct Method.**
23. Lamperti: **Stochastic Processes: A Survey of the Mathematical Theory.**
24. Grenander: **Pattern Analysis: Lectures in Pattern Theory, Vol. II.**
25. Davies: **Integral Transforms and Their Applications.**
26. Kushner/Clark: **Stochastic Approximation Methods for Constrained and Unconstrained Systems.**
27. de Boor: **A Practical Guide to Splines.**
28. Keilson: **Markov Chain Models—Rarity and Exponentiality.**
29. de Veubeke: **A Course in Elasticity.**
30. Sniatycki: **Geometric Quantization and Quantum Mechanics.**
31. Reid: **Sturmian Theory for Ordinary Differential Equations.**
32. Meis/Markowitz: **Numerical Solution of Partial Differential Equations.**
33. Grenander: **Regular Structures: Lectures in Pattern Theory, Vol. III.**
34. Kevorkian/Cole: **Perturbation Methods in Applied Mathematics.**
35. Carr: **Applications of Centre Manifold Theory.**

(continued on inside back cover)

K.W. Chang
F.A. Howes

Nonlinear Singular Perturbation Phenomena: Theory and Applications

Springer-Verlag
New York Berlin Heidelberg Tokyo

K.W. Chang
Department of Mathematics
University of Calgary
Calgary, Alberta
Canada T2N 1N4

F.A. Howes
Department of Mathematics
University of California
Davis, California 95616
U.S.A.

AMS Classification: 34D15, 34D20, 34EXX, 35B20, 35B25, 35F99, 35G99

Library of Congress Cataloging in Publication Data
Chang, K.W.
Nonlinear singular perturbation phenomena.
(Applied mathematical sciences ; v. 56)
Bibliography: p.
Includes indexes.
1. Boundary value problems–Numerical solutions.
2. Singular perturbations (Mathematics) I. Howes, Frederick A. II. Title. III. Series:
Applied mathematical sciences (Springer-Verlag New York Inc.); v. 56.
QA1.A647 vol. 56 [QA379] 510 s [515.3′5] 84-14014

With 12 Illustrations

© 1984 by Springer-Verlag New York Inc.
All rights reserved. No part of this book may be translated or reproduced in any form without written permission from Springer-Verlag, 175 Fifth Avenue, New York, New York, 10010, U.S.A.

Printed and bound by R.R. Donnelley & Sons, Harrisonburg, Virginia.
Printed in the United States of America.

9 8 7 6 5 4 3 2 1

ISBN 0-387-96066-X Springer-Verlag New York Berlin Heidelberg Tokyo
ISBN 3-540-96066-X Springer-Verlag Berlin Heidelberg New York Tokyo

Preface

Our purpose in writing this monograph is twofold. On the one hand, we want to collect in one place many of the recent results on the existence and asymptotic behavior of solutions of certain classes of singularly perturbed nonlinear boundary value problems. On the other, we hope to raise along the way a number of questions for further study, mostly questions we ourselves are unable to answer. The presentation involves a study of both scalar and vector boundary value problems for ordinary differential equations, by means of the consistent use of differential inequality techniques. Our results for scalar boundary value problems obeying some type of maximum principle are fairly complete; however, we have been unable to treat, under any circumstances, problems involving "resonant" behavior. The linear theory for such problems is incredibly complicated already, and at the present time there appears to be little hope for any kind of general nonlinear theory. Our results for vector boundary value problems, even those admitting higher dimensional maximum principles in the form of invariant regions, are also far from complete. We offer them with some trepidation, in the hope that they may stimulate further work in this challenging and important area of differential equations.

The research summarized here has been made possible by the support over the years of the National Science Foundation and the National Science and Engineering Research Council. We offer each agency our sincerest thanks for their generosity and consideration. We also wish to thank our colleagues and students who have shared their knowledge of and curiosity about singular perturbation theory with us, especially Bob O'Malley,

Adelaida Vasil'eva and Wolfgang Wasow. This monograph is but a small token of our appreciation of their friendship and support.

K. W. Chang
Calgary

F. A. Howes
Davis

Contents

Chapter I
Introduction

We are mainly interested in quasilinear and nonlinear boundary value problems, to which some well-known methods, such as the methods of matched asymptotic expansions and two-variable expansions are not immediately applicable. For example, let us consider the following boundary value problem (cf. O'Malley [75], Chapter 5)

$$\varepsilon y'' = y'^2, \quad 0 < t < 1, \tag{A}$$

$$y(0,\varepsilon) = 1, y(1,\varepsilon) = 0. \tag{B}$$

In general, it is not obvious that such a nonlinear boundary value problem will have a solution in $[0,1]$ for all sufficiently small values of ε. However, in this case, we can obtain by quadratures the following exact solution in $[0,1]$

$$y(t,\varepsilon) = -\varepsilon \ln[t + e^{-1/\varepsilon}(1-t)]$$

which is defined for all positive values of ε.

An important feature of this solution $y(t,\varepsilon)$ is that, as a function of (t,ε), it behaves nonuniformly as t and ε approach 0, that is,

$$\lim_{\varepsilon\to 0^+} y(t,\varepsilon) = 0 \quad \text{for each fixed} \quad t > 0 \tag{1.1}$$

but

$$\lim_{t\to 0^+} y(t,\varepsilon) = 1 \quad \text{for each fixed} \quad \varepsilon > 0. \tag{1.2}$$

For decreasing values of ε, the solutions $y(t,\varepsilon)$ are as shown in Figure 1.1.

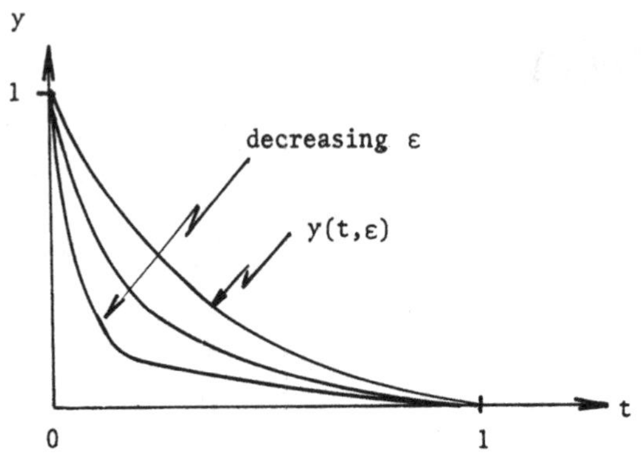

Figure 1.1

From the graph or from the relations (1.1), (1.2), it is clear that, as $\varepsilon \to 0^+$, the solution $y(t,\varepsilon)$ approaches 0 uniformly with respect to t on each closed subinterval of $(0,1]$ but not on the whole interval $[0,1]$. Note that $u(t) \equiv 0$ (which is the limit of the solution $y(t,\varepsilon)$ on $[\delta,1]$, for fixed $\delta > 0$) turns out to be the solution of the corresponding reduced equation

$$u'^2 = 0$$

of the original equation (A) satisfying the right-hand boundary condition (B).

In order to illustrate the difficulties associated with the application of the methods of matched and two-variable asymptotic expansions, we assume, for the moment, that the exact solution of problem (A), (B) is not known, and so we proceed formally. The fundamental hypothesis of these methods is that the solution of (A), (B) can be described by two power series in ε, known as the inner and outer expansions. The outer expansion represents the solution away from regions of nonuniform behavior and is simply a power series in ε whose coefficients are functions of t. On the other hand, the coefficients of the inner expansion are functions not only of t but also of a "stretched" variable $\tau = \frac{t}{\varepsilon}$ which can be arbitrarily large as $\varepsilon \to 0^+$ for a certain range of t. The variable τ (cf. [55], Chapter 2; [75], Chapter 1) can be regarded as a rescaling parameter which has the effect of enlarging the region of nonuniform behavior. To fix these ideas, let us examine first the outer expansion $y_0(t,\varepsilon)$, that is, we substitute

$$y_0(t,\varepsilon) = u_0(t) + \varepsilon u_1(t) + \dots = \sum_0^\infty \varepsilon^n u_n(t)$$

into the differential equation $\varepsilon y'' = y'^2$ and equate coefficients of like powers of ε. The first two terms of y_0 are easily shown to satisfy the equations

$$u_0'^2 = 0 \tag{1.3}$$

and

$$u_1'' = u_1'^2. \tag{1.4}$$

The solutions of (1.3) are $u_0(t)$ = constant, while the problem of solving (1.4) is essentially equivalent to solving the original problem (A), (B). We note that $u_1(t)$ = constant is one family of solutions of (1.4). If we require that $y_0(t,\varepsilon)$ satisfy one of the original boundary conditions (B), then the obvious choices for the functions u_0 and u_1 are

$$u_0(t) \equiv 0, \quad u_1(t) \equiv 0 \quad (\text{if } y_0(1,\varepsilon) = 0)$$

$$u_0(t) \equiv 1, \quad u_1(t) \equiv 0 \quad (\text{if } y_0(0,\varepsilon) = 1).$$

The first choice $y_0(1,\varepsilon) = 0$ turns out to be the proper requirement here (cf. [75], Chapter 5) and it can be motivated geometrically as follows. If the outer expansion $y_0(t,\varepsilon)$ were to satisfy the second choice $y_0(0,\varepsilon) = 1$, then we would anticipate that

$$\lim_{\varepsilon\to 0^+} y(t,\varepsilon) = 1, \quad \text{for } t \text{ in } [0,1-\delta], \text{ where } \delta > 0.$$

In a neighborhood of $t = 1$, $y(t,\varepsilon)$ must decrease rapidly from one to zero in order to satisfy the other boundary condition $y(1,\varepsilon) = 0$. Since the equation (A) requires that y' never change sign, we must have $y'' < 0$. This however is impossible since $\varepsilon y'' = y'^2 > 0$. Consequently, we must select the outer expansion which satisfies the boundary condition $y_0(1,\varepsilon) = 0$ at $t = 1$.

In this case, the outer expansion $y_0(t,\varepsilon)$ should be identically zero (up to terms of order ε^2); however, the function (cf. [75], Chapter 5)

$$\tilde{y}_0(t,\varepsilon) = -\varepsilon \ln t$$

is also an outer expansion and

$$\tilde{y}_0(1,\varepsilon) = 0.$$

This function has a singularity at $t = 0$ for $\varepsilon > 0$ and so one might be tempted to reject it out of hand as an approximation to y on $(0,1]$. The surprising fact is that this function $\tilde{y}_0$ *is* the outer expansion, as follows from the exact solution. Indeed, the singularity of $\tilde{y}_0$ at $t = 0$ is precisely what is needed to cancel the singularity of the inner expansion there.

The construction of the inner expansion of the solution y is equally fraught with difficulties, as it is not obvious at all what the correct stretched variable τ should be. To see this we could make the standard change of variable (cf. [55], Chapter 2; [75], Chapter 5)

$$\tau = \frac{t}{\phi(\varepsilon)} \quad \text{for} \quad \phi(\varepsilon) > 0 \quad \text{and} \quad \phi(\varepsilon) \to 0^+ \quad \text{as} \quad \varepsilon \to 0^+,$$

and attempt to determine the asymptotic character of $\phi(\varepsilon)$ by comparing the terms of the transformed differential equation. Clearly $\varepsilon y'' = y'^2$ is equivalent to

$$\frac{\varepsilon}{\phi^2(\varepsilon)} \frac{d^2y}{d\tau^2} = \frac{1}{\phi^2(\varepsilon)} \left(\frac{dy}{d\tau}\right)^2,$$

and so the change of variable accomplishes nothing. Using various devices, O'Malley ([75], Chapter 5) is able to construct an inner expansion which in fact has a singularity at $t = 0$ that just cancels the singularity of the function y there. His methods are nevertheless not obvious a priori, and it is quite conceivable that more complicated problems of the form (A), (B) involving y'^2-nonlinearities could not be solved to such a degree.

We can, however, solve this particular problem using our method, but rather than discussing this method now, we will defer it until Chapter V, when we deal with a much larger class of related boundary value problems.

Notes and Remarks.

1.1. The methods of matched asymptotic expansions and two-variable expansions have been very successful in solving a variety of difficult problems in engineering and applied science, and they continue to be two of the most powerful weapons in the arsenal of applied mathematicians. Our discussion of the nonlinear problem (A), (B) (and indeed, the point of view taken in this monograph), is not meant to denigrate, in any way, the utility and importance of these stalwarts of asymptotic analysis. Rather we wish to study certain types of boundary value problems, for which precise results on the existence and asymptotic behavior of solutions can be derived from mathematical analysis.

1.2. Alternative approaches to many of the problems discussed in this book can be found in the monographs of Wasow ([93], Chapter 10), Vasil'eva and Butuzov [88], O'Malley [75], Habets [28], Habets and Laloy [31], Eckhaus [21] and Kevorkian and Cole [55], and in the survey articles of Vasil'eva [87], Erdélyi [23], O'Malley [73], Carrier [8] and Wasow [94]. In addition, these monographs and papers contain a wealth of references.

Chapter II
A'priori Bounds and Existence Theorems

§2.1. Scalar Boundary Value Problems

Before discussing in detail the various classes of singularly perturbed boundary value problems, let us give an outline of the principal method of proof that we will use throughout. This method employs the theory of differential inequalities which was developed by M. Nagumo [66] and later refined by Jackson [49]. It enables one to prove the existence of a solution, and at the same time, to estimate this solution in terms of the solutions of appropriate inequalities. Such an approach has been found to be very useful in a number of different applications (see, for example, [5] and [83]). It will be seen that for the general classes of problems which we will study in later chapters, this inequality technique leads elegantly (and easily) to some fairly general results about existence of solutions and their asymptotic behavior. Many results which have been obtained over the years by a variety of methods can now be obtained by this method, which we hope will also very clearly reveal the fundamental asymptotic processes at work.

Consider first the general Dirichlet problem

$$x'' = f(t,x,x'), \quad a < t < b,$$
$$x(a) = A, \quad x(b) = B, \tag{DP}$$

in which f is a continuous function on $[a,b] \times \mathbb{R}^2$. The differential inequality approach of Nagumo is based on the observation that if there exist smooth (say twice continuously differentiable or $C^{(2)}$-) functions $\alpha(t)$ and $\beta(t)$ possessing the following properties:

$$\begin{cases} \alpha(t) \le \beta(t) \\ \alpha(a) \le A \le \beta(a), \quad \alpha(b) \le B \le \beta(b) \\ \alpha'' \ge f(t,\alpha,\alpha'), \quad \beta'' \le f(t,\beta,\beta'), \end{cases} \tag{P}$$

then the problem (DP) has a solution $x = x(t)$ of class $C^{(2)}([a,b])$ such that $\alpha(t) \le x(t) \le \beta(t)$ for t in $[a,b]$, provided that f does not grow "too fast" as a function of x'. More precisely, it is sufficient to require that f satisfies what is known as a generalized Nagumo condition with respect to α and β (cf. [36]). This simply means that every solution $x = x(t)$ of $x'' = f(t,x,x')$ satisfying $\alpha(t) \le x(t) \le \beta(t)$ on a subinterval $J \subset [a,b]$ has a bounded derivative, that is, there exists a constant $N = N(\alpha,\beta)$ such that $|x'(t)| \le N$ on J. The most common type of Nagumo condition is the following:

$$f(t,x,z) = O(|z|^2) \quad \text{as} \quad |z| \to \infty \quad \text{for all} \quad (t,x) \quad \text{in} \quad [a,b] \times [\alpha,\beta].$$

This was originally given by Nagumo [66] himself. Clearly, if

$$x'' = f(t,x,x') = O(|x'|^2)$$

and if $\alpha(t) \le x \le \beta(t)$, then $|x'| \le N(\alpha,\beta)$. In summary then, we have

Theorem 2.1. Assume that there exist bounding functions $\alpha(t)$ and $\beta(t)$ with the properties (P), and assume that the function f satisfies a generalized Nagumo condition with respect to α and β. Then the Dirichlet problem (DP) has a solution $x = x(t)$ of class $C^{(2)}([a,b])$ satisfying $\alpha(t) \le x(t) \le \beta(t)$ for t in $[a,b]$.

Thus the task of estimating solutions of (DP) is reduced to the task of constructing sufficiently sharp bounding functions $\alpha(t)$ and $\beta(t)$. In this regard, we note (cf. [31]) that it is possible to obtain the same result as in Theorem 2.1 if the bounding functions α and β are only piecewise $-C^{(2)}$ on $[a,b]$, that is, if there is a partition $\{t_i\}$ of $[a,b]$ with $a = t_0 < t_1 < t_2 < \dots < t_n = b$, such that on each subinterval $[t_{i-1},t_i]$, α and β are twice continuously differentiable. At the partition points t_{i-1} and t_i, the derivatives are the right-hand and left-hand derivatives, respectively. We must of course assume that on each subinterval (t_{i-1},t_i), $\alpha'' \ge f(t,\alpha,\alpha')$, $\beta'' \le f(t,\beta,\beta')$, and furthermore that for each t in (a,b), $\alpha'(t^-) \le \alpha'(t^+)$ and $\beta'(t^-) \ge \beta'(t^+)$. Then there is the following companion result.

Theorem 2.2. Assume that there exist piecewise $-C^{(2)}$ bounding functions α and β with the stated properties, and assume that the function f

satisfies a generalized Nagumo condition with respect to α and β. Then the conclusion of Theorem 2.1 follows.

This result allows the bounding functions to have certain types of "corners". It follows from the observation that if $\{\alpha_1,\ldots,\alpha_m\}[\{\beta_1,\ldots,\beta_M\}]$ are lower [upper] functions, then so is $\max\{\alpha_1,\ldots,\alpha_m\}[\min\{\beta_1,\ldots,\beta_M\}]$. We will need this fact in our discussion of interior layer behavior associated with some singular perturbation problems.

Let us now consider a boundary value problem with more general boundary conditions of the form

$$\begin{aligned} &x'' = f(t,x,x'), \quad a < t < b, \\ &p_1x(a) - p_2x'(a) = A, \quad q_1x(b) + q_2x'(b) = B, \end{aligned} \tag{RP}$$

where the constants p_i, q_i satisfy $p_2,q_2 \geq 0$, $p_1^2 + p_2^2 > 0$ and $q_1^2 + q_2^2 > 0$. Note that if $p_2 = q_2 = 0$ and $p_1 = q_1 = 1$, then the problem (RP) reduces to (DP), and so we are really interested in the case when $p_2 + q_2 > 0$. It turns out fortunately that the Nagumo theory for (DP) can be extended with the obvious modifications to the problem (RP), as was observed by Heidel [36]. That is, if there exist piecewise $-C^{(2)}$ bounding functions α and β $(\alpha \leq \beta)$ which satisfy the above differential inequalities and the boundary inequalities $p_1\alpha(a) - p_2\alpha'(a) \leq A \leq p_1\beta(a) - p_2\beta'(a)$, $q_1\alpha(b) + q_2\alpha'(b) \leq B \leq q_1\beta(b) + q_2\beta'(b)$, then the problem (RP) has a solution $x = x(t)$ such that $\alpha(t) \leq x(t) \leq \beta(t)$ for t in $[a,b]$, provided that f satisfies a generalized Nagumo condition with respect to α and β. For later reference, we call this result

<u>Theorem 2.3</u>. Assume that there exist piecewise $-C^{(2)}$ bounding functions α and β with the stated properties, and assume that f satisfies a generalized Nagumo condition with respect to α and β. Then the Robin problem (RP) has a solution $x = x(t)$ of class $C^{(2)}([a,b])$ with $\alpha(t) \leq x(t) \leq \beta(t)$ for t in $[a,b]$.

In studying singularly perturbed boundary value problems one is frequently interested in obtaining theorems which guarantee a priori the existence of solutions and give an estimation of the location of the solutions. The most common results of this kind for both perturbed and unperturbed problems are obtained using some maximum principle argument in which the solution is estimated throughout its interval of existence in terms of its values on the boundary of the interval. The remainder of this section is concerned basically with existence and estimation results which follow either directly or indirectly (that is, after a change of variable) from the one-dimensional maximum principle, and its generalizations as embodied in Theorems 2.1 - 2.3.

We consider first the Dirichlet problem

$$\varepsilon y'' = f(t,y,y'), \quad a < t < b, \tag{2.1}$$

$$y(a,\varepsilon) = A, \quad y(b,\varepsilon) = B, \tag{2.2}$$

in which $f = O(|y'|^2)$ as $|y'| \to \infty$, that is, for (t,y) in compact subsets of $[a,b] \times \mathbb{R}$, $f(t,y,y') = O(|y'|^2)$ as $|y'| \to \infty$. The next result is a direct application of the maximum principle (cf. [81]).

Lemma 2.1. Assume that the function f is continuous with respect to t,y,y' and of class $C^{(1)}$ with respect to y for (t,y,y') in $[a,b] \times \mathbb{R}^2$. Assume also that there is a positive constant m such that $f_y(t,y,0) \geq m > 0$ for (t,y) in $[a,b] \times \mathbb{R}$. Then for each $\varepsilon > 0$, the problem (2.1), (2.2) has a unique solution $y = y(t,\varepsilon)$ in $[a,b]$ satisfying

$$|y(t,\varepsilon)| \leq M/m,$$

where

$$M = \max\{\max_{[a,b]} |f(t,0,0)|, m|A|, m|B|\}.$$

Proof: Define for t in $[a,b]$

$$\alpha(t) = -M/m \quad \text{and} \quad \beta(t) = M/m.$$

Then $\alpha \leq \beta$, $\alpha(a) \leq A \leq \beta(a)$ and $\alpha(b) \leq B \leq \beta(b)$. To obtain the differential inequalities $\varepsilon\alpha'' \geq f(t,\alpha,\alpha')$, $\varepsilon\beta'' \leq f(t,\beta,\beta')$, we note that by Taylor's Theorem

$$f(t,\alpha,0) = f(t,0,0) + f_y(t,\xi,0)\alpha$$

where ξ, $\alpha < \xi < 0$, is an intermediate point, and so

$$f(t,\alpha,0) \leq |f(t,0,0)| + m\alpha \leq M + m(-M/m) \leq 0 = \varepsilon\alpha''.$$

Similarly, for some intermediate point η, $0 < \eta < \beta$,

$$f(t,\beta,0) = f(t,0,0) + f_y(t,\eta,0) \geq -M + m(M/m) \geq 0 = \varepsilon\beta''.$$

It follows from Theorem 2.1 that for each $\varepsilon > 0$ the problem (2.1), (2.2) has a solution $y(t,\varepsilon)$ on $[a,b]$ satisfying

$$-M/m \leq y(t,\varepsilon) \leq M/m.$$

The uniqueness of the solution follows from the maximum principle.

If we assume that $A \geq 0$, $B \geq 0$ and

$-M \leq f(t,0,0) \leq 0$ in $[a,b]$,

then by the proof of Lemma 2.1, we will obtain the following more precise estimate of the solution:

$0 \leq y(t,\varepsilon) \leq M/m$.

Similarly, if we assume that $A \leq 0$, $B \leq 0$ and $M \geq f(t,0,0) \geq 0$ in $[a,b]$, then the solution of (2.1), (2.2) satisfies $-M/m \leq y(t,\varepsilon) \leq 0$. These results suggest the following modifications of Lemma 2.1.

Lemma 2.2. Assume that the function f is continuous with respect to t,y,y' and of class $C^{(n)}$ $(n \geq 2)$ with respect to y for (t,y,y') in $[a,b] \times \mathbb{R}^2$. Assume also that $A \geq 0$, $B \geq 0$, $f(t,0,0) \leq 0$, and that there is a positive constant m such that $\partial_y^j f(t,0,0) \geq 0$ for $1 \leq j \leq n-1$ and $\partial_y^n f(t,y,0) \geq m > 0$ for (t,y) in $[a,b] \times \mathbb{R}$. Then for each $\varepsilon > 0$, the problem (2.1), (2.2) has a solution $y = y(t,\varepsilon)$ in $[a,b]$ satisfying

$$0 \leq y(t,\varepsilon) \leq (n!m^{-1}M)^{1/n},$$

where

$$M = \max\{\max_{[a,b]} |f(t,0,0)|,\ (m|A|/n!)^n,\ (m|B|/n!)^n\}.$$

Remark. We use the notation $\partial_y^n f$ for $\partial^n f/\partial y^n$.

Proof: Define for t in $[a,b]$

$$\alpha(t) \equiv 0 \quad \text{and} \quad \beta(t) = (n!m^{-1}M)^{1/n}.$$

Clearly, $\alpha \leq \beta$, $\alpha(a) \leq A \leq \beta(a)$, $\alpha(b) \leq B \leq \beta(b)$, and $\varepsilon\alpha'' \geq f(t,\alpha,\alpha')$ by virtue of our assumptions on A, B and f. Finally, we see that $\varepsilon\beta'' \leq f(t,\beta,\beta')$ since

$$\begin{aligned} f(t,\beta,0) &= f(t,0,0) + \sum_{j=1}^{n-1} \partial_y^j f(t,0,0)\beta^j/j! \\ &\quad + \partial_y^n f(t,\eta,0)\beta^n/n! \\ &\geq -M + (m/n!)n!\ M/m \\ &\geq 0. \end{aligned}$$

The conclusion of Lemma 2.2 follows by virtue of Theorem 2.1.

If we make the change of dependent variable $y \to -y$ and apply Lemma 2.2 to the transformed problem, we can obtain an analogous result for the case $A \le 0$, $B \le 0$ and $f(t,0,0) \ge 0$. This is given in the next lemma.

Lemma 2.3. Assume that the function f is continuous with respect to t,y,y' and of class $C^{(n)}$ $(n \ge 2)$ with respect to y for (t,y,y') in $[a,b] \times \mathbb{R}^2$. Assume also that $A \le 0$, $B \le 0$, $f(t,0,0) \ge 0$, and that there is a positive constant m such that $\partial_y^{j_0(j_e)} f(t,0,0) \ge 0$ (≤ 0) for $1 \le j_0(j_e) \le n-1$ and $\partial_y^n f(t,y,0) \ge m > 0$ $(\le -m < 0)$ if n is odd (even), for (t,y) in $[a,b] \times \mathbb{R}$. (Here $j_0(j_e)$ denotes an odd (even) integer.) Then for each $\varepsilon > 0$, the problem (2.1), (2.2) has a solution $y = y(t,\varepsilon)$ in $[a,b]$ satisfying

$$-(n!m^{-1}M)^{1/n} \le y(t,\varepsilon) \le 0,$$

where M is as defined in the conclusion of Lemma 2.2.

In the previous lemmas, we imposed strong conditions on the partial derivatives of f with respect to y. In the lemma below we will relax our conditions on the partial derivatives of f with respect to y, but we will impose a condition on the partial derivative $f_{y'}$, and thereby obtain virtually the same result as in Lemma 2.1.

Lemma 2.4. Assume that the function f is continuous with respect to t,y,y' and of class $C^{(1)}$ with respect to y,y' for (t,y,y') in $[a,b] \times \mathbb{R}^2$. Assume also that there are positive constants ℓ and k such that $|f_y(t,y,0)| \le \ell$ for (t,y) in $[a,b] \times \mathbb{R}$, and $|f_{y'}(t,y,y')| \ge k > 0$ for (t,y,y') in $[a,b] \times \mathbb{R}^2$. Then for $0 < \varepsilon < k^2/4\ell$, the problem (2.1), (2.2) has a solution $y = y(t,\varepsilon)$ in $[a,b]$ satisfying

$$|y(t,\varepsilon)| \le \gamma\ell^{-1}(2e^{\lambda(t-b)} - 1) \quad (\text{if } f_{y'} < 0)$$

or

$$|y(t,\varepsilon)| \le \gamma\ell^{-1}(2e^{\lambda(a-t)} - 1) \quad (\text{if } f_{y'} > 0).$$

Here $\lambda = -\ell k^{-1} + O(\varepsilon)$ is a negative root of the polynomial $\varepsilon\lambda^2 + k\lambda + \ell$ for $0 < \varepsilon \le \varepsilon_0 < k^2/4\ell$, $\gamma = \max\{\max_{[a,b]} |f(t,0,0)|, \ell|A|(2e^{\lambda(a-b)}-1), \ell|B|\}$ if $f_{y'} < 0$, and $\gamma = \max\{\max_{[a,b]} |f(t,0,0)|, \ell|A|, \ell|B|(2e^{\lambda(a-b)} - 1)$ if $f_{y'} > 0$.

Proof: Suppose for definiteness that $f_{y'} \le -k < 0$. Define for $0 < \varepsilon \le \varepsilon_0$ and t in $[a,b]$

$$\alpha(t,\varepsilon) = -\gamma\ell^{-1}(2e^{\lambda(t-b)} - 1)$$

and

$$\beta(t,\varepsilon) = \gamma\ell^{-1}(2e^{\lambda(t-b)} - 1).$$

Then, clearly $\alpha \leq \beta$, $\alpha(a) \leq A \leq \beta(a)$ and $\alpha(b) \leq B \leq \beta(b)$ by our choice of γ. We will only verify that $\varepsilon\alpha'' \geq f(t,\alpha,\alpha')$ in (a,b), as the verification that $\varepsilon\beta'' \leq f(t,\beta,\beta')$ in (a,b) proceeds analogously. Differentiating and substituting, we have

$$\begin{aligned}
\varepsilon\alpha'' - f(t,\alpha,\alpha') &= \varepsilon\alpha'' - f(t,0,0) - \{f(t,\alpha,0) - f(t,0,0)\} \\
&\qquad - \{f(t,\alpha,\alpha') - f(t,\alpha,0)\} \\
&= \varepsilon\alpha'' - f(t,0,0) - f_y(t,\xi,0)\alpha - f_{y'}(t,\alpha,\eta)\alpha' \\
&\geq -\varepsilon\lambda^2\gamma\ell^{-1}2e^{\lambda(t-b)} - \gamma - \ell\gamma\ell^{-1}2e^{\lambda(t-b)} + \gamma \\
&\qquad - k\lambda\gamma\ell^{-1}2e^{\lambda(t-b)} \\
&= 0,
\end{aligned}$$

since $\varepsilon\lambda^2 + k\lambda + \ell = 0$. The conclusion of the lemma in the case that $f_{y'} \leq -k < 0$ now follows from Theorem 2.1. If $f_{y'} \geq k > 0$ then we define for $0 < \varepsilon \leq \varepsilon_0$ and t in $[a,b]$

$$\alpha(t,\varepsilon) = -\gamma\ell^{-1}(2e^{\lambda(a-t)} - 1), \quad \beta(t,\varepsilon) = -\alpha(t,\varepsilon),$$

and proceed as above.

We consider finally a sufficient condition for the existence of a solution of the problem (2.1), (2.2) which includes the assumptions $f_y > 0$ and $|f_{y'}| > 0$ of Lemmas 2.1 and 2.3, respectively. The next lemma is due essentially to van Harten [33].

<u>Lemma 2.5</u>. Assume that the function f is continuous with respect to t,y,y' and of class $C^{(1)}$ with respect to y,y' for (t,y,y') in $[a,b] \times \mathbb{R}^2$. Assume also that there are a constant ν and a positive constant m such that $f_y(t,y,0) + \nu f_{y'}(t,y,y') \geq m + \varepsilon\nu^2$ for (t,y,y') in $[a,b] \times \mathbb{R}^2$ and $0 < \varepsilon \leq \varepsilon_1$. Then for $0 < \varepsilon \leq \varepsilon_1$, the problem (2.1), (2.2) has a solution $y = y(t,\varepsilon)$ in $[a,b]$ satisfying

$$|y(t,\varepsilon)| \leq m^{-1}Ne^{\nu t},$$

where

$$N = \max\{\max_{[a,b]} |f(t,0,0)e^{-\nu t}|, \quad m|A|e^{-a\nu}, \quad m|B|e^{-b\nu}\}.$$

<u>Proof:</u> The lemma follows by making the change of variable $y = ze^{\nu t}$ and

applying Lemma 2.1 to the resulting problem for z, namely $\varepsilon z'' = F(t,z,z',\varepsilon)$, $z(a,\varepsilon) = Ae^{-a\nu}$, $z(b,\varepsilon) = Be^{-b\nu}$, where $F(t,z,z',\varepsilon) = f(t,0,0)e^{-\nu t} + \{f_y(t,\xi,0) + \nu f_{y'}(t,ze^{\nu t},\eta) - \varepsilon\nu^2\}z + \{f_{y'}(t,ze^{\nu t},\eta) - 2\varepsilon\nu\}z'$.

We note that if $f_y \geq m > 0$, then we may take $\nu = 0$ in Lemma 2.5, while if $|f_{y'}| \geq k > 0$, then we may take ν such that $\operatorname{sgn}\nu = \operatorname{sgn} f_{y'}$ and $|\nu k| > \ell$ for $\ell = \sup|f_y(t,y,0)|$, in order to derive the results of Lemmas 2.1 and 2.4, respectively.

Lastly we wish to point out that analogous results hold for the Robin problem

$$\varepsilon y'' = f(t,y,y'), \quad a < t < b, \tag{2.3}$$

$$p_1 y(a,\varepsilon) - p_2 y'(a,\varepsilon) = A, \quad q_1 y(b,\varepsilon) + q_2 y'(b,\varepsilon) = B. \tag{2.4}$$

In particular, if $f(t,y,y') = O(|y'|^2)$ as $|y'| \to \infty$ for (t,y) in compact subsets of $[a,b] \times \mathbb{R}$, and if $p_1 = q_1 = 1$, then Lemmas 2.1 - 2.4 hold verbatim for (2.3), (2.4), as the reader can easily verify.

§2.2. Vector Boundary Value Problems

Analogous results also hold for vector boundary value problems. The existence and comparison theorems for vector problems can be regarded as higher dimensional forms of Nagumo's scalar theory. Unfortunately, however, the assumptions which are imposed for vector problems are more difficult to verify in practice. This is due, on the one hand, to our limited experience in treating boundary value problems for systems of differential equations. On the other hand, systems of differential equations are inherently more complicated than scalar equations, and so at best, we can only hope to mimic the scalar theory. The results which follow are taken mostly from the papers of Kelley [52], [53], although much of the early work was done by Hartman, [34], [35; Chapter 12] and others (see [5] and [83] for further references).

Consider then the boundary value problem

$$\begin{aligned} &\underset{\sim}{x}'' = \underset{\sim}{F}(t,\underset{\sim}{x},\underset{\sim}{x}'), \quad a < t < b, \\ &\underset{\sim}{x}(a) = \underset{\sim}{A}, \quad \underset{\sim}{x}(b) = \underset{\sim}{B}, \end{aligned} \tag{$\underset{\sim}{\mathrm{DP}}$}$$

where $\underset{\sim}{x}$, $\underset{\sim}{A}$ and $\underset{\sim}{B}$ are vectors in $\mathbb{R}^N$ and $\underset{\sim}{F} \equiv (F_1,\ldots,F_N)^T$ is an N-vector function which is continuous on $[a,b] \times \mathbb{R}^{2N}$. It turns out that the scalar Nagumo theory can be extended to ($\underset{\sim}{\mathrm{DP}}$), provided that the vector function $\underset{\sim}{F}$ satisfies a growth condition (Nagumo condition) with respect to $\underset{\sim}{x}'$. In this vector setting, we say that $\underset{\sim}{F}$ satisfies a Nagumo condition if it satisfies one of the following two conditions, for $(t,\underset{\sim}{x})$ in compact subsets of $[a,b] \times \mathbb{R}^N$ and for all $\underset{\sim}{z}$ in $\mathbb{R}^N$ (cf. [53]):

(1) There exist positive, nondecreasing, continuous functions ϕ_i on $(0,\infty)$ such that each component F_i, $i = 1,\ldots,N$ of $\underset{\sim}{F}$ satisfies

$$|F_i(t,\underset{\sim}{x},\underset{\sim}{z})| \leq \phi_i(|z_i|)$$

and

$$\int^{\infty} s/\phi_i(s)ds = \infty;$$

(2) There exists a positive, nondecreasing, continuous function ϕ on $(0,\infty)$ such that

$$||\underset{\sim}{F}(t,\underset{\sim}{x},\underset{\sim}{z})|| \leq \phi(||\underset{\sim}{z}||)$$

and

$$s^2/\phi(s) \to \infty \quad \text{as} \quad s \to \infty.$$

Here and below, $||\cdot||$ is the usual Euclidean norm, that is, $||\underset{\sim}{x}|| \equiv (\underset{\sim}{x}^T\underset{\sim}{x})^{1/2}$ for any $\underset{\sim}{x} = (x_1,\ldots,x_N)^T$ in $\mathbb{R}^N$. It is clear that if $\underset{\sim}{F}$ is independent of $\underset{\sim}{x}'$, then $\underset{\sim}{F}$ satisfies both types of Nagumo condition; while, if $\underset{\sim}{F}$ depends linearly on $\underset{\sim}{x}'$, then in general, we can only say that $\underset{\sim}{F}$ satisfies condition (2). Clearly condition (1) is satisfied if each component F_i of $\underset{\sim}{F}$ depends only on x_i', and if $F_i(t,\underset{\sim}{x},z_i) = O(|z_i|^2)$ as $|z_i| \to \infty$, for $(t,\underset{\sim}{x})$ in compact subsets of $[a,b] \times \mathbb{R}^N$.

We now extend the scalar notion of a bounding function by supposing that there exist M scalar functions $\rho_\ell = \rho_\ell(t,\underset{\sim}{x})$, $\ell = 1,\ldots,M$, of class $C^{(2)}([a,b] \times \mathbb{R}^N)$ satisfying

$$\rho_\ell'' \geq 0 \quad \text{whenever} \quad \rho_\ell = 0 \quad \text{and} \quad \rho_\ell' = 0. \tag{2.5}$$

Here, the derivatives

$$\rho_\ell' \equiv \frac{\partial \rho_\ell}{\partial t} + [\text{grad } \rho_\ell]^T \underset{\sim}{x}'$$

and

$$\rho_\ell'' \equiv \frac{\partial^2 \rho_\ell}{\partial t^2} + [2 \text{ grad } \partial\rho_\ell/\partial t]^T \underset{\sim}{x}' + \underset{\sim}{x}'^T H \underset{\sim}{x}' + [\text{grad } \rho_\ell]^T \underset{\sim}{F}(t,\underset{\sim}{x},\underset{\sim}{x}')$$

are taken with respect to the solution of $(\underset{\sim}{D}P)$, and H is the Hessian matrix of ρ_ℓ with respect to $\underset{\sim}{x}$, (that is, H is the Jacobian matrix that represents the derivative of grad ρ_ℓ). Next, we define the following region

$$I \equiv \{(t,\underset{\sim}{x}) \text{ in } [a,b] \times \mathbb{R}^n\colon \rho_\ell(t,\underset{\sim}{x}) \leq 0 \text{ for } \ell = 1,\ldots,M\},$$

which plays an important role in the following result, a vector analog of Theorem 2.1, which is taken from Kelley [53].

Theorem 2.4. Assume that there exist M functions $\rho_\ell = \rho_\ell(t,\underset{\sim}{x})$, $\ell = 1,\ldots,M$, satisfying (2.5) and that the function $\underset{\sim}{F} = \underset{\sim}{F}(t,\underset{\sim}{x},\underset{\sim}{x}')$ satisfies a Nagumo condition in the domain $I \times \mathbb{R}^N$. Assume also that the initial and terminal pairs $(a,\underset{\sim}{A})$, $(b,\underset{\sim}{B})$ belong to I and can be joined by a smooth path in I. Then the Dirichlet problem $(\underset{\sim}{D}P)$ has a solution $\underset{\sim}{x} = \underset{\sim}{x}(t)$ of class $C^{(2)}([a,b])$ such that $\rho_\ell(t,\underset{\sim}{x}(t)) \leq 0$ on $[a,b]$, for $\ell = 1,\ldots,M$.

The conclusion of the theorem is that $(\underset{\sim}{D}P)$ has a solution $\underset{\sim}{x}(t)$ with $(t,\underset{\sim}{x}(t))$ in the region I; hence, we call I an invariant region for $(\underset{\sim}{D}P)$. Thus the description of the nature of the solution of $(\underset{\sim}{D}P)$ is dependent on the appropriate comparison functions ρ_ℓ.

Suppose first we use a single comparison function $\rho = \rho(t,\underset{\sim}{x}) = ||\underset{\sim}{x}|| - \gamma(t)$, where $||\underset{\sim}{A}|| \leq \gamma(a)$ and $||\underset{\sim}{B}|| \leq \gamma(b)$. If ρ satisfies (2.5), then we obtain $||\underset{\sim}{x}|| \leq \gamma(t)$ on $[a,b]$, which gives the magnitude of the bound on the norm $||\underset{\sim}{x}||$. To ensure that (2.5) is satisfied, we proceed to determine the scalar function $\gamma(t)$. A calculation shows that $\operatorname{grad} \rho = \underset{\sim}{x}/||\underset{\sim}{x}||$ and $\operatorname{grad}(\partial\rho/\partial t) \equiv \underset{\sim}{0}$; furthermore, as the norm $||\cdot||$ is a convex function, it follows that the Hessian matrix H of $||\underset{\sim}{x}||$ (and hence, of ρ) is nonnegative definite, that is, $\underset{\sim}{x}^T H \underset{\sim}{x} \geq 0$ for all $\underset{\sim}{x}$ in $\mathbb{R}^N$. The inequality (2.5) then reduces to

$$\rho'' \geq -\gamma'' + (\underset{\sim}{x}^T/||\underset{\sim}{x}||)\underset{\sim}{F}(t,\underset{\sim}{x},\underset{\sim}{x}') \geq 0$$

or

$$\gamma''(t) \leq (\underset{\sim}{x}^T/||\underset{\sim}{x}||)\underset{\sim}{F}(t,\underset{\sim}{x},\underset{\sim}{x}'),$$

whenever $\gamma(t) = ||\underset{\sim}{x}||$ and $\gamma'(t) = (\underset{\sim}{x}^T/||\underset{\sim}{x}||)\underset{\sim}{x}'$. If appropriate assumptions are imposed on $\underset{\sim}{F}$, then $\gamma(t)$ can be determined.

Suppose, on the other hand, we wish to obtain bounds on the individual components of the solution of $(\underset{\sim}{D}P)$; then we obviously need more than one comparison function. To this end, we define the $2N$ functions

$$\rho_\ell(t,\underset{\sim}{x}) = x_\ell - \beta_\ell(t) \quad \text{and} \quad \rho_{\ell+N}(t,\underset{\sim}{x}) = -x_\ell + \alpha_\ell(t),$$

where $\alpha_\ell \leq \beta_\ell$, for $\ell = 1,\ldots,N$. If these functions satisfy (2.5), as well as the boundary inequalities

$$\alpha_\ell(a) \leq A_\ell \leq \beta_\ell(a), \quad \alpha_\ell(b) \leq B_\ell \leq \beta_\ell(b),$$

then we obtain two-sided bounds of the form

$$\alpha_\ell(t) \leq x_\ell(t) \leq \beta_\ell(t)$$

for $\ell = 1,\ldots,N$, on $[a,b]$ (cf. Theorem 2.1). The reader can verify that the functions above will satisfy (2.5), if α_ℓ and β_ℓ satisfy, respectively, the inequalities

$$\alpha_\ell'' \geq F_\ell(t,\underset{\sim}{x},\underset{\sim}{x}') \quad \text{when} \quad x_\ell = \alpha_\ell(t), \quad x_\ell' = \alpha_\ell'(t)$$

and

$$\beta_\ell'' \leq F_\ell(t,\underset{\sim}{x},\underset{\sim}{x}') \quad \text{when} \quad x_\ell = \beta_\ell(t), \quad x_\ell' = \beta_\ell'(t),$$

for all x_i $(i \neq \ell)$ in $[\alpha_i(t),\beta_i(t)]$ and all x_i' $(i \neq \ell)$ in $\mathbb{R}$.

By means of different types of invariant regions in Theorem 2.4, we can prove *a priori* existence and comparison results for the general singularly perturbed vector problem

$$\varepsilon\underset{\sim}{y}'' = \underset{\sim}{F}(t,\underset{\sim}{y},\underset{\sim}{y}'), \quad a < t < b,$$

$$\underset{\sim}{y}(a,\varepsilon) = \underset{\sim}{A}, \quad \underset{\sim}{y}(b,\varepsilon) = \underset{\sim}{B},$$

and these can be regarded as analogous to Lemmas 2.1 - 2.5. The analysis is rather tedious, although straightforward, and we leave the precise formulation of these results to the reader.

The last existence and comparison theorem of this chapter deals with the following vector Robin problem

$$\underset{\sim}{x}'' = \underset{\sim}{H}(t,\underset{\sim}{x}), \quad a < t < b,$$
$$P\underset{\sim}{x}(a) - \underset{\sim}{x}'(a) = \underset{\sim}{A}, \quad Q\underset{\sim}{x}(b) + \underset{\sim}{x}'(b) = \underset{\sim}{B}, \tag{$\underset{\sim}{\mathrm{RP}}$}$$

where $\underset{\sim}{x}$, $\underset{\sim}{A}$ and $\underset{\sim}{B}$ are N-vectors, P, Q are constant $(N \times N)$-matrices, and $\underset{\sim}{H}$ is an N-vector function defined and continuous on $[a,b] \times \mathbb{R}^N$. (We have limited ourselves to discussing only systems of singularly perturbed Robin problems whose right-hand sides do not depend on any derivatives, but of course we could have discussed more general systems.) If the matrices P and Q are positive semidefinite in the sense that there exist nonnegative scalars p and q such that $\underset{\sim}{x}^T P\underset{\sim}{x} \geq p||\underset{\sim}{x}||^2$ and $\underset{\sim}{x}^T Q\underset{\sim}{x} \geq q||\underset{\sim}{x}||^2$, for any $\underset{\sim}{x}$ in $\mathbb{R}^N$, then it is possible to prove a result analogous to Theorem 2.3. Since we will only seek bounds on the norm of a solution of $(\underset{\sim}{\mathrm{RP}})$, we call a set $I \equiv \{(t,\underset{\sim}{x})$ in $[a,b] \times \mathbb{R}^N$: $\rho(t,\underset{\sim}{x}) \leq 0\}$ an invariant region for $(\underset{\sim}{\mathrm{RP}})$ if the scalar function ρ has the following three properties:

(1) ρ is of class $C^{(2)}(I)$;

(2) $p\rho(a,\underset{\sim}{x}(a)) - \rho'(a,\underset{\sim}{x}(a)) \leq 0$,
$q\rho(b,\underset{\sim}{x}(b)) + \rho'(b,\underset{\sim}{x}(b)) \leq 0$;

(3) $\rho'' \geq 0$ in I whenever $\rho = 0$ and $\rho' = 0$.

(The functions ρ', ρ'' are as defined in (2.5), with the function $\underset{\sim}{H}(t,\underset{\sim}{x})$ in place of $\underset{\sim}{F}$.) Then we have the following result (cf. [52], [57]).

Theorem 2.5. Assume that there exists an invariant region I for (RP). Then the Robin problem (RP) has a solution $\underset{\sim}{x} = \underset{\sim}{x}(t)$ of class $C^{(2)}([a,b])$ such that $\rho(t,\underset{\sim}{x}(t)) \leq 0$ in $[a,b]$.

This theorem will be used in Chapter VII to estimate the norms of solutions of singularly perturbed Robin problems. There we will show, under appropriate assumptions, how to construct functions of the form $\rho(t,\underset{\sim}{x},\varepsilon) = ||\underset{\sim}{x}|| - \gamma(t,\varepsilon)$ which satisfy the conditions (1) - (3).

Notes and Remarks

2.1. Most of the differential inequality and invariant region results quoted in this chapter can be found in the monographs of Bernfeld and Lakshmikantham [5] and Schröder [83]. These works also contain many additional references to the relevant literature, as well as instructive illustrations of other applications of inequality techniques.

2.2. M. Nagumo in a paper [67] published in 1939 was the first mathematician to apply differential inequalities in the study of a singular perturbation problem, namely the initial value problem $\varepsilon y'' = f(t,y,y',\varepsilon)$, $0 < t \leq T < \infty$, $y(0,\varepsilon)$, $y'(0,\varepsilon)$ prescribed. This paper was overlooked until the Soviet mathematician N. I. Bris in [7] used Nagumo's results to study the singularly perturbed boundary problem $\varepsilon y'' = f(t,y,y',\varepsilon)$, $p_1y(a,\varepsilon) - p_2y'(a,\varepsilon)$, $q_1y(b,\varepsilon) + q_2y'(b,\varepsilon)$ prescribed, by means of a shooting technique. Most, if not all, of the present monograph is based on these two seminal papers; the reader would do well to consult them or the brief survey [45].

2.3. It is possible to extend Theorem 2.5 to the Robin problem for the differential equation $\underset{\sim}{x}'' = \underset{\sim}{F}(t,\underset{\sim}{x},\underset{\sim}{x}')$ and to more general invariant regions of the type considered in Theorem 2.4; cf. [52] or [57].

Chapter III
Semilinear Singular Perturbation Problems

§3.1. The Dirichlet Problem: Boundary Layer Phenomena

We consider first the semilinear Dirichlet problem

$$\varepsilon y'' = h(t,y), \quad a < t < b,$$
$$y(a,\varepsilon) = A, \quad y(b,\varepsilon) = B, \tag{DP_1}$$

where ε is a small positive parameter and prime denotes differentiation with respect to t. Some natural questions to ask regarding this problem are: Does the problem have a solution for all small values of ε? Once the existence of a solution has been established, how does the solution behave as $\varepsilon \to 0^+$?

The answers to these questions depend greatly on the function h (and also on the boundary values A, B, if h is a nonlinear function), as we shall see by examining two simple linear equations in $(0,1)$

$$\varepsilon y'' = y \tag{E_1}$$

and

$$\varepsilon y'' = -y \tag{E_2}$$

subject to the boundary conditions

$$y(0,\varepsilon) = 1, \quad y(1,\varepsilon) = 2.$$

The general solution of (E_1) is

$$y = y_1(t,\varepsilon) = c_1 \exp(t/\sqrt{\varepsilon}) + c_2 \exp(-t/\sqrt{\varepsilon})$$

and the general solution of (E_2) is

$$y = y_2(t,\varepsilon) = c_1 \cos(t/\sqrt{\varepsilon}) + c_2 \sin(t/\sqrt{\varepsilon}),$$

where c_1 and c_2 are arbitrary constants. Using the boundary conditions to determine these constants, we find that the solutions are, respectively,

$$y_1(t,\varepsilon) = (e^{-1/\sqrt{\varepsilon}} - e^{1/\sqrt{\varepsilon}})^{-1}\{(e^{-1/\sqrt{\varepsilon}} - 2)e^{t/\sqrt{\varepsilon}} + (2-e^{1/\sqrt{\varepsilon}})e^{-t/\sqrt{\varepsilon}}\}$$

$$\sim 2e^{-(1-t)/\sqrt{\varepsilon}} + e^{-t/\sqrt{\varepsilon}}$$

and

$$y_2(t,\varepsilon) = \cos(t/\sqrt{\varepsilon}) + \{2 - \cos(1/\sqrt{\varepsilon})\}\sin(t/\sqrt{\varepsilon})\{\sin(1/\sqrt{\varepsilon})\}^{-1}.$$

Let us examine these solutions more closely. For (E_1), the solution y_1 is defined for all $\varepsilon > 0$ and, moreover,

$$\lim_{\varepsilon\to 0^+} y_1(t,\varepsilon) = 0 \quad \text{for} \quad \delta \leq t \leq 1-\delta, \tag{3.1}$$

where δ is a fixed constant in $(0,1)$. The function y_1 attains the limiting value 0 nonuniformly in the neighborhoods of $t = 0$ and $t = 1$ in the following sense:

$$\lim_{\varepsilon\to 0^+}\lim_{t\to 0^+} y_1(t,\varepsilon) = 1 \neq 0 = \lim_{t\to 0^+}\lim_{\varepsilon\to 0^+} y_1(t,\varepsilon)$$

and

$$\lim_{\varepsilon\to 0^+}\lim_{t\to 1^-} y_1(t,\varepsilon) = 2 \neq 0 = \lim_{t\to 1^-}\lim_{\varepsilon\to 0^+} y_1(t,\varepsilon)$$

(cf. Figure 3.1). We note that by setting $\varepsilon = 0$ in (E_1) we obtain the reduced equation $0 = h(t,u)$ whose solution is $u \equiv 0$.

In the case of (E_2), first of all we see that the solution $y_2(t,\varepsilon)$ is only defined if $\sin(1/\sqrt{\varepsilon}) \neq 0$, that is, if $\sqrt{\varepsilon} \neq (n\pi)^{-1}$, for $n = 1, 2, \ldots$. Thus, for these particular values of ε, the problem (E_2) has no solution. Suppose then ε is very small and $\sqrt{\varepsilon} \neq (n\pi)^{-1}$. The function y_2 is a linear combination of two oscillatory functions of arbitrarily large arguments, and therefore it is densely oscillatory with period $\sqrt{\varepsilon}$ and with bounded amplitude (cf. Figure 3.2). Clearly, it is impossible for y_2 to satisfy a limiting relation such as (3.1) above, even though $u \equiv 0$ is again the solution of the reduced equation obtained from (E_2) by setting $\varepsilon = 0$. What sets these two problems apart is of course the difference in the sign of the coefficient of y.

Let us now consider the third example in which h is a nonlinear function of y in $(0,1)$,

$$\varepsilon y'' = y^2, \quad y(0,\varepsilon) = A, \quad y(1,\varepsilon) = B. \tag{E_3}$$

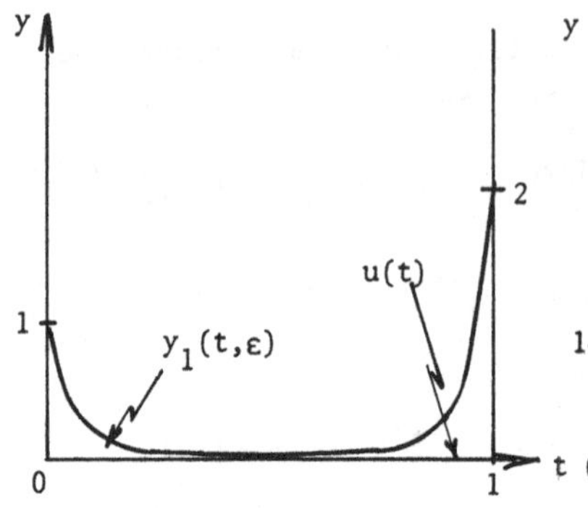

Figure 3.1

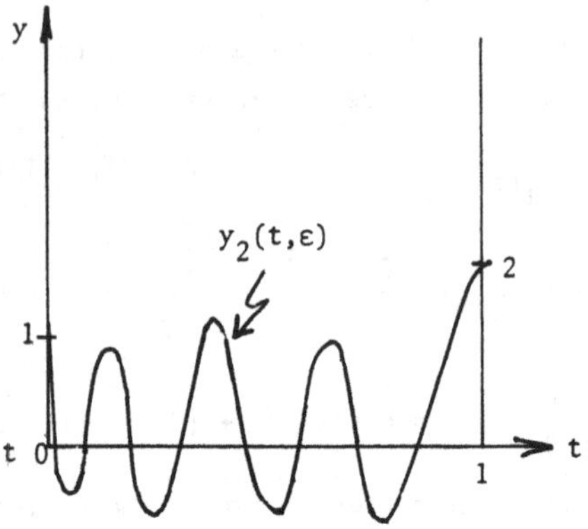

Figure 3.2

This example illustrates how the boundary values A and B can affect the existence and the behavior of solutions as $\varepsilon \to 0^+$. From our consideration of the example (E_1), we expect that for sufficiently small $\varepsilon > 0$, the solution of (E_3) must remain close to zero, the unique solution of the reduced equation $y^2 = 0$, except possibly near the endpoints $t = 0$ and $t = 1$. However, in the neighborhood of $t = 0$ and $t = 1$, the solution must be convex (i.e., $y'' > 0$), as dictated by the differential equation $y'' = \varepsilon^{-1}y^2 > 0$. Thus, if either the boundary value A or B is negative, we would have $y'' < 0$ near $t = 0$ or $t = 1$, respectively; see Figure 3.3. In these cases, the problem has *no* solution for sufficiently small values of ε. On the other hand, if both A and B are nonnegative, we have $y'' > 0$ near $t = 0$ and $t = 1$, as Figure 3.4 shows. For such values of A and B, the problem (E_3) has a solution $y = y_3(t,\varepsilon)$ for all sufficiently small values of ε, and

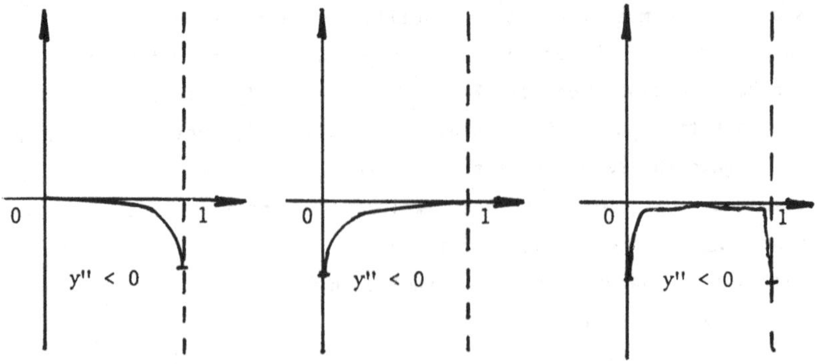

Figure 3.3

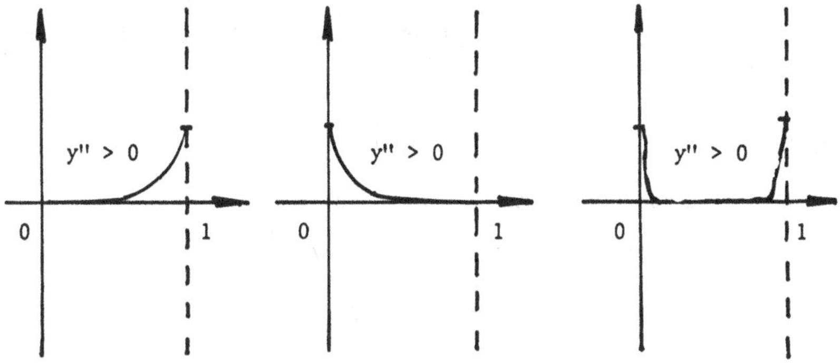

Figure 3.4

$$\lim_{\varepsilon\to 0^+} y_3(t,\varepsilon) = 0 \quad \text{for} \quad \delta \le t \le 1-\delta,$$

where δ is a fixed constant in $(0,1)$.

It is interesting to note that the exact solution $y_3(t,\varepsilon)$ (for $A,B \ge 0$) can be obtained in terms of elliptic integrals [76], from which we see that

$$y(t,\varepsilon) \sim A[1 + t(A/6\varepsilon)^{1/2}]^{-2} + B[1 + (1-t)(B/6\varepsilon)^{1/2}]^{-2}$$

in $[0,1]$.

However, by simply observing that the solution must satisfy $y'' \ge 0$, we are able to deduce the sign restrictions on the allowable boundary values A,B.

If we apply this observation and reasoning to the same problem for the differential equation $\varepsilon y'' = y^3$, we conclude that the problem has a solution for any value of A and B.

Rather than dealing with specific cases, we study now the existence of solutions $y = y(t,\varepsilon)$ of the general problem (DP_1) (and related ones) which behave like the solutions of (E_1) and (E_3) in the sense that

$$\lim_{\varepsilon\to 0^+} y(t,\varepsilon) = u(t) \quad \text{in each closed subinterval of } (a,b), \tag{3.2}$$

where $u = u(t)$ is a certain solution of the reduced equation

$$h(t,u) = 0, \quad a \le t \le b. \tag{R_1}$$

If the relation (3.2) holds, we can say that the reduced solution $u(t)$ is "stable" with respect to the original solution $y(t,\varepsilon)$. To be precise, we wish to give explicit definitions of stability for the solution $u(t)$. Let $u = u(t)$ be a solution of (R_1) which is continuous in $[a,b]$, and let us define the following three domains $\mathcal{D}_0(u)$, $\mathcal{D}_1(u)$, $\mathcal{D}_2(u)$:

$$\mathcal{D}_0(u) = \{(t,y):\ a \le t \le b,\ \ |y - u(t)| \le d(t)\},$$

where $d(t)$ is a positive continuous function such that

$$d(t) \equiv |A - u(a)| + \delta, \quad \text{for } a \le t \le a + \delta/2$$

$$d(t) \equiv \delta, \quad \text{for } a + \delta \le t \le b - \delta$$

and

$$d(t) \equiv |B - u(b)| + \delta, \quad \text{for } b - \delta/2 \le t \le b.$$

Here $\delta > 0$ is a small constant.

If $A \ge u(a)$ and $B \ge u(b)$, we define

$$\mathcal{D}_1(u) = \{(t,y):\ a \le t \le b,\ \ 0 \le y - u(t) \le d(t)\},$$

and if $A \le u(a)$ and $B \le u(b)$, we define

$$\mathcal{D}_2(u) = \{(t,y):\ a \le t \le b,\ \ -d(t) \le y - u(t) \le 0\},$$

where $d(t)$ is as above.

In the following definitions of stability for the solution $u(t)$, we assume that the function $h(t,y)$ has the stated number of continuous partial derivatives with respect to y in $\mathcal{D}_i(u)$, $i = 0,1$ or 2, and that $q \ge 0$ and $n \ge 2$ are integers.

<u>Definition 3.1</u>. The function $u = u(t)$ is said to be (I_q)-stable in $[a,b]$ if there exists a positive constant m such that

$$\partial_y^j h(t,u(t)) \equiv 0 \quad \text{for} \quad a \le t \le b \quad \text{and} \quad 0 \le j \le 2q,$$

and

$$\partial_y^{2q+1} h(t,y) \ge m > 0 \quad \text{in} \quad \mathcal{D}_0(u).$$

<u>Definition 3.2</u>. The function $u = u(t)$ is said to be $(\tilde{I}_q)$-stable in $[a,b]$ if there exists a positive constant m such that

$$\partial_y^j h(t,u(t)) \equiv 0 \quad \text{for} \quad a \le t \le b \quad \text{and} \quad 0 \le j \le 2q,$$

$$\partial_y^{2q+1} h(t,u(t)) \ge m > 0 \quad \text{for} \quad a \le t \le b,$$

$$(A-u(a))\partial_y^{2q+2} h(t,y) \ge 0 \quad \text{in} \quad \mathcal{D}_0(u) \cap ([a,a+\delta] \times \mathbb{R}),$$

and

$$(B-u(b))\partial_y^{2q+2}h(t,y) \geq 0 \quad \text{in} \quad \mathcal{D}_0(u) \cap ([b-\delta,b] \times \mathbb{R}).$$

The above definitions are motivated by boundary value problems for the differential equation $\varepsilon y'' = y^{2q+1}$ (cf. (E_1) above).

Definition 3.3. The function $u = u(t)$ is said to be (II_n)-stable in $[a,b]$ if $u(a) \leq A$, $u(b) \leq B$ and if there exists a positive constant m such that

$$\partial_y^j h(t,u(t)) \geq 0 \quad \text{for} \quad a \leq t \leq b \quad \text{and} \quad 1 \leq j \leq n-1,$$

and

$$\partial_y^n h(t,y) \geq m > 0 \quad \text{in} \quad \mathcal{D}_1(u).$$

Definition 3.4. The function $u = u(t)$ is said to be $(\tilde{II}_n)$-stable in $[a,b]$ if $u(a) \leq A$, $u(b) \leq B$ and if there exists a positive constant m such that

$$\partial_y^j h(t,u(t)) \geq 0 \quad \text{for} \quad a \leq t \leq b \quad \text{and} \quad 1 \leq j \leq n-1,$$

$$\partial_y^n h(t,u(t)) \geq m > 0 \quad \text{for} \quad a \leq t \leq b,$$

and

$$\partial_y^{n+1} h(t,y) \geq 0 \quad \text{in} \quad \mathcal{D}_1(u) \cap [([a,a+\delta) \cup (b-\delta,b]) \times \mathbb{R}].$$

The above definitions are motivated by boundary value problems for the differential equation $\varepsilon y'' = y^{2n}$ (cf. (E_3) above).

Definition 3.5. The function $u = u(t)$ is said to be (III_n)-stable in $[a,b]$ if $u(a) \geq A$, $u(b) \geq B$ and if there exists a positive constant m such that

$$\partial_y^{j_0(j_e)} h(t,u(t)) \geq 0 \;\; (\leq 0) \quad \text{for} \quad a \leq t \leq b \quad \text{and} \quad 1 \leq j_0,\ j_e \leq n-1,$$

where $j_0(j_e)$ denotes an odd (even) integer, and

$$\partial_y^n h(t,y) \leq -m < 0 \;\; (\geq m > 0) \quad \text{in} \quad \mathcal{D}_2(u), \quad \text{if } n \text{ is even (odd)}.$$

Definition 3.6. The function $u = u(t)$ is said to be $(\tilde{III}_n)$-stable in $[a,b]$ if $u(a) \geq A$, $u(b) \geq B$ and if there exists a positive constant m such that

$$\partial_y^{j_0(j_e)} h(t,u(t)) \geq 0 \;\; (\leq 0) \quad \text{for} \quad a \leq t \leq b \quad \text{and} \quad 1 \leq j_0,\ j_e \leq n-1,$$

$$\partial_y^n h(t,u(t)) \leq -m < 0 \quad (\geq m > 0) \quad \text{for} \quad a \leq t \leq b \quad \text{if} \quad n \quad \text{is} \quad \text{even (odd)},$$

and

$$\partial_y^n h(t,y) \geq 0 \quad (\leq 0) \quad \text{in} \quad \mathcal{D}_2(u) \cap [([a,a+\delta) \cap (b-\delta,b]) \times \mathbb{R}],$$

is n is even (odd).

The last two definitions are motivated by boundary value problems for the differential equation $\varepsilon y'' = -y^{2n}$.

With these definitions of stability we proceed now to discuss the Dirichlet problem (DP_1). We remark that the constant c in each theorem of of this chapter is a known positive constant depending on the reduced path under consideration.

Theorem 3.1. Assume that the reduced equation (R_1) has an (I_q)- or ($\tilde{I}_q$)-stable solution $u = u(t)$ of class $C^{(2)}([a,b])$. Then there exists an $\varepsilon_0 > 0$ such that for $0 < \varepsilon \leq \varepsilon_0$ the problem (DP_1) has a solution $y = y(t,\varepsilon)$ for t in $[a,b]$ which satisfies

$$|y(t,\varepsilon)-u(t)| \leq w_L(t,\varepsilon) + w_R(t,\varepsilon) + c\varepsilon^{1/(2q+1)},$$

where

$$w_L(t,\varepsilon) = |A-u(a)| \exp[-(m\varepsilon^{-1})^{1/2}(t-a)] \quad \text{if} \quad q = 0,$$

$$w_L(t,\varepsilon) = |A-u(a)| \; (1+\sigma|A-u(a)|^q\varepsilon^{-1/2}(t-a))^{-1/q} \quad \text{if} \quad q \geq 1,$$

$$w_R(t,\varepsilon) = |B-u(b)| \exp[-(m\varepsilon^{-1})^{1/2}(b-t)] \quad \text{if} \quad q = 0,$$

and

$$w_R(t,\varepsilon) = |B-u(b)| \; (1+\sigma|B-u(b)|^q\varepsilon^{-1/2}(b-t))^{-1/q} \quad \text{if} \quad q \geq 1.$$

Here

$$\sigma = m^{1/2}q[(q+1)(2q+1)!]^{-1/2}$$

and c is some positive constant.

Proof: The theorem follows from Theorem 2.1 of Chapter II, if we can exhibit, by construction, the existence of the lower and the upper bounding functions $\alpha(t,\varepsilon)$ and $\beta(t,\varepsilon)$ with the required properties.

Since, by assumption, $\partial_y^{2q+1}h(t,y) \geq m > 0$, we must have $h(t,y) \sim my^{2q+1}/(2q+1)!$, and we are led to consider the differential equation

$$\varepsilon w'' = \frac{m}{(2q+1)!} w^{2q+1}. \tag{3.3}$$

Indeed, the function $w_L(t,\varepsilon)$ is nonnegative and is the solution of (3.3) such that $w_L(a,\varepsilon) = |A-u(a)|$, $w_L'(a,\varepsilon) = -[\frac{m}{\varepsilon(q+1)(2q+1)!}]^{1/2}|A-u(a)|^{q+1}$. The solution decreases to the right. Similarly, the function $w_R(t,\varepsilon) \geq 0$ is the solution of (3.3) such that $w_R(b,\varepsilon) = |B-u(b)|$, $w_R'(b,\varepsilon) = [\frac{m}{\varepsilon(q+1)(2q+1)!}]^{1/2}|B-u(b)|^{q+1}$. It decreases to the left.

We now define, for t in $[a,b]$ and $\varepsilon > 0$, the functions

$$\alpha(t,\varepsilon) = u(t) - w_L(t,\varepsilon) - w_R(t,\varepsilon) - \Gamma(\varepsilon),$$

$$\beta(t,\varepsilon) = u(t) + w_L(t,\varepsilon) + w_R(t,\varepsilon) + \Gamma(\varepsilon).$$

Here $\Gamma(\varepsilon) = (\varepsilon\gamma/m)^{1/(2q+1)}$, where γ is a positive constant which will be specified later.

It is obvious that the functions α,β have the following properties: $\alpha \leq \beta$, $\alpha(a,\varepsilon) \leq A \leq \beta(a,\varepsilon)$ and $\alpha(b,\varepsilon) \leq B \leq \beta(b,\varepsilon)$. It is just as easy to prove that $\varepsilon\alpha'' \geq h(t,\alpha)$ and $\varepsilon\beta'' \leq h(t,\beta)$ in (a,b) for some suitable choice of γ. We treat the case that $u(t)$ is (I_q)-stable and consider $\alpha(t,\varepsilon)$. (The verification for $\beta(t,\varepsilon)$ follows by symmetry.) From Taylor's Theorem and the hypothesis that $u(t)$ is (I_q)-stable, we have

$$\begin{aligned} h(t,\alpha(t,\varepsilon)) &= h(t,\alpha(t,\varepsilon)) - h(t,u(t)) \\ &= \sum_{n=1}^{2q} \frac{1}{n!}\,\partial_y^n h(t,u(t))\,[\alpha(t,\varepsilon) - u(t)]^n \\ &\qquad + \frac{1}{(2q+1)!}\,\partial_y^{2q+1} h(t,\xi(t))[\alpha(t,\varepsilon) - u(t)]^{2q+1} \\ &= -\frac{1}{(2q+1)!}\,\partial_y^{2q+1} h(t,\xi(t))\,(w_L+w_R+\Gamma)^{2q+1}, \end{aligned}$$

where $(t,\xi(t))$ is some intermediate "point" between $(t,\alpha(t,\varepsilon))$ and $(t,u(t))$, which lies in $\mathcal{D}_0(u)$ for sufficiently small ε, say $0 < \varepsilon \leq \varepsilon_0$. Since w_L, w_R and Γ are all positive functions, we have

$$-h(t,\alpha(t,\varepsilon)) \geq \frac{m}{(2q+1)!}\,(w_L^{2q+1} + w_R^{2q+1} + \Gamma^{2q+1}),$$

and so

$$\begin{aligned} \varepsilon\alpha'' - h(t,\alpha(t,\varepsilon)) &\geq \varepsilon u'' - \varepsilon w_L'' - \varepsilon w_R'' + \frac{m}{(2q+1)!}(w_L^{2q+1}+w_R^{2q+1}+\Gamma^{2q+1}) \\ &\geq -\varepsilon|u''| + \frac{\varepsilon\gamma}{(2q+1)!}, \end{aligned}$$

by the definitions of w_L, w_R, and Γ. Thus, by choosing $\gamma \geq |u''|(2q+1)!$, we obtain $\varepsilon\alpha'' \geq h(t,\alpha)$.

The case that $u(t)$ is $(\tilde{I}_q)$-stable can be treated analogously, by carrying out the Taylor expansion to $(2q+2)$ terms; the details are left to the reader.

If the reduced equation (R_1) has a (II_n)- or $(\tilde{II}_n)$-stable solution, we have the following result.

Theorem 3.2. Assume that the reduced equation (R_1) has a (II_n)- or $(\tilde{II}_n)$-stable solution $u = u(t)$ of class $C^{(2)}([a,b])$ such that $u(a) \leq A$, $u(b) \leq B$ and $u'' \geq 0$ in (a,b). Then there exists an $\varepsilon_0 > 0$ such that for $0 < \varepsilon \leq \varepsilon_0$ the problem (DP_1) has a solution $y = y(t,\varepsilon)$ in $[a,b]$ which satisfies

$$0 \leq y(t,\varepsilon) - u(t) \leq w_L(t,\varepsilon) + w_R(t,\varepsilon) + c\varepsilon^{1/n},$$

where

$$w_L(t,\varepsilon) = (A-u(a))(1+\sigma(A-u(a))^{1/2(n-1)}\varepsilon^{-1/2}(t-a))^{-2/(n-1)}$$

and

$$w_R(t,\varepsilon) = (B-u(b))(1+\sigma(B-u(b))^{1/2(n-1)}\varepsilon^{-1/2}(b-t))^{-2/(n-1)}.$$

Here

$$\sigma = (n-1)(m/2(m+1)!)^{1/2}$$

and c is some positive constant.

Proof: The proof of Theorem 3.2 follows in much the same manner the proof of the previous theorem, once we note that $w_L > 0$ is now the solution of the differential equation $\varepsilon w'' = \frac{m}{n!} w^n$ which satisfies $w_L(a,\varepsilon) = A - u(a)$ and $w_L'(a,\varepsilon) = -(2m/\varepsilon(n+1)!)^{1/2}(A-u(a))^{1/2(n+1)}$, and that $w_R > 0$ is the solution satisfying $w_R(b,\varepsilon) = B - u(b)$ and $w_R'(b,\varepsilon) = (2m/\varepsilon(n+1)!)^{1/2}(B-u(b))^{1/2(n+1)}$. We then define

$$\alpha(t,\varepsilon) = u(t),$$

$$\beta(t,\varepsilon) = u(t) + w_L(t,\varepsilon) + w_R(t,\varepsilon) + (\varepsilon\gamma m^{-1})^{1/n},$$

for $\gamma \geq |u''|n!$, and proceed as above. We leave details to the reader except to note that the convexity of u implies that $\varepsilon\alpha'' - h(t,\alpha) = \varepsilon u'' - h(t,u) = \varepsilon u'' \geq 0$.

The next theorem is the analog of Theorem 3.2 when the solution of the reduced equation is (III_n)- or $(\tilde{III}_n)$-stable. It can be proved easily by making the change of variable $y \to -y$ and immediately applying Theorem 3.2.

Theorem 3.3. Assume that the reduced equation (R_1) has a (III_n)- or $(\tilde{II}I_n)$-stable solution $u = u(t)$ of class $C^{(2)}([a,b])$ such that $u(a) \geq A$, $u(b) \geq B$ and $u'' \leq 0$ in (a,b). Then there exists an $\varepsilon_0 > 0$ such that for $0 < \varepsilon \leq \varepsilon_0$ the problem (DP_1) has a solution $y = y(t,\varepsilon)$ in $[a,b]$ which satisfies

$$-w_L(t,\varepsilon) - w_R(t,\varepsilon) - c\varepsilon^{1/n} \leq y(t,\varepsilon) - u(t) \leq 0,$$

where w_L, w_R and c are the same as in Theorem 3.2.

§3.2. Robin Problems: Boundary Layer Phenomena

We turn now to a consideration of the Robin problems

$$\begin{aligned} &\varepsilon y'' = h(t,y), \quad a < t < b, \\ &y(a,\varepsilon) - p_1 y'(a,\varepsilon) = A, \quad y(b,\varepsilon) = B, \end{aligned} \tag{RP_1}$$

and

$$\begin{aligned} &\varepsilon y'' = h(t,y), \quad a < t < b, \\ &y(a,\varepsilon) - p_1 y'(a,\varepsilon) = A, \quad y(b,\varepsilon) + p_2 y'(b,\varepsilon) = B, \end{aligned} \tag{RP_2}$$

where p_1 and p_2 are positive constants. (The Robin problem with the boundary conditions $y(a,\varepsilon)$ and $y(b,\varepsilon) + p_2 y'(b,\varepsilon)$ prescribed, can be handled by making the change of variable $t \to a + b - t$ in (RP_1).) In order to motivate our results, we examine briefly the behavior of solutions of the linear equation $\varepsilon y'' = y$ satisfying the above types of boundary conditions. Consider first the problem

$$\begin{aligned} &\varepsilon y'' = y, \quad 0 < t < 1, \\ &y(0,\varepsilon) - y'(0,\varepsilon) = 1, \quad y(1,\varepsilon) = 1. \end{aligned} \tag{E_4}$$

The exact solution is easily found to be

$$y(t,\varepsilon) = \Delta^{-1}\left\{(1-\varepsilon^{-1/2}-e^{\varepsilon^{-1/2}}) \exp[-t\varepsilon^{-1/2}] + (e^{-\varepsilon^{-1/2}}-1-\varepsilon^{-1/2})\exp[t\varepsilon^{-1/2}]\right\},$$

where

$$\Delta = e^{-\varepsilon^{-1/2}}(1-\varepsilon^{-1/2}) - e^{\varepsilon^{-1/2}}(1+\varepsilon^{-1/2}).$$

Clearly

$$y(t,\varepsilon) \sim \varepsilon^{1/2} \exp[-t\varepsilon^{-1/2}] + \exp[-(1-t)\varepsilon^{-1/2}],$$

and it follows that

$$\lim_{\varepsilon\to 0^+} y(t,\varepsilon) = 0 \quad \text{for} \quad 0 \le t \le 1 - \delta, \tag{3.4}$$

where δ, $0<\delta<1$, is a fixed constant. Thus in contrast to the limiting relation (3.1), the solution of (E_4) is uniformly close to its limiting value of 0 near $t = 0$. Similarly the solution $y = y(t,\varepsilon)$ of the problem

$$\varepsilon y'' = y, \quad 0 < t < 1,$$
$$y(0,\varepsilon) - y'(0,\varepsilon) = 1, \quad y(1,\varepsilon) + y'(1,\varepsilon) = 1, \tag{E_5}$$

can be obtained explicitly and it satisfies

$$\lim_{\varepsilon\to 0^+} y(t,\varepsilon) = 0 \quad \text{for} \quad 0 \le t \le 1. \tag{3.5}$$

The difference between the relations (3.1) and (3.4), (3.5) stems from the fact that the boundary conditions in (E_4) and (E_5) prescribe $y'(0,\varepsilon)$ or $y'(1,\varepsilon)$ to be uniformly bounded in ε, and so the solutions of these Robin problems cannot behave like the solution of (E_1) near $t = 0$ and/or $t = 1$.

In terms of the Robin problem (RP_1), a solution $u = u(t)$ of the reduced equation (R_1) is said to be (I_q)-, $(\tilde{I}_q)$-, (II_n)-, $(\tilde{II}_n)$-, (III_n)- or $(\tilde{III}_n)$-stable if it is so stable in the sense of Definitions 3.1 - 3.6, respectively, with $d(t)$ satisfying $d(t) \equiv \delta$ in $[a, b - \delta]$ and $d(t) \equiv |B-u(b)| + \delta$ in $[b-\delta/2, b]$.

The following theorems are analogs of Theorems 3.1 - 3.3.

<u>Theorem 3.4</u>. Assume that the reduced equation (R_1) has an (I_q)- or $(\tilde{I}_q)$-stable solution $u = u(t)$ of class $C^{(2)}([a,b])$. Then there exists an $\varepsilon_0 > 0$ such that for $0 < \varepsilon \le \varepsilon_0$ the problem (RP_1) has a solution $y = y(t,\varepsilon)$ in $[a,b]$ which satisfies

$$|y(t,\varepsilon)-u(t)| \le v_L(t,\varepsilon) + w_R(t,\varepsilon) + c\varepsilon^{1/(2q+1)}.$$

Here

$$v_L(t,\varepsilon) = (\varepsilon/mp_1^2)^{1/2}|A-u(a) + p_1u'(a)| \exp[-(m/\varepsilon)^{1/2}(t-a] \quad \text{if } q = 0,$$

$$v_L(t,\varepsilon) = \sigma[1 + q(\varepsilon(2q+2)!/2m)^{-1/2}\sigma^q(t-a)]^{-1/q} \quad \text{if } q \ge 1,$$

where

$$\sigma^{q+1} = \{\varepsilon(2q+2)!/2mp_1^2\}^{1/2}|A-u(a) + p_1u'(a)|,$$

w_R is as given in Theorem 3.1, and c is some positive constant.

Proof: The proof of this result is not that much different from the proof of Theorem 3.1 which deals with the Dirichlet problem (DP_1). Indeed, the function $v_L > 0$ is the decaying solution of the differential equation $\varepsilon z'' = \frac{m}{(2q+1)!} z^{2q+1}$ which satisfies $v_L'(a,\varepsilon) = -|A-u(a) + p_1u'(a)|/p_1$ and $v_L(a,\varepsilon) = \sigma$. Thus, for t in $[a,b]$ and $\varepsilon > 0$ we define

$$\alpha(t,\varepsilon) = u(t) - v_L(t,\varepsilon) - w_R(t,\varepsilon) - \Gamma(\varepsilon),$$

$$\beta(t,\varepsilon) = u(t) + v_L(t,\varepsilon) + w_R(t,\varepsilon) + \Gamma(\varepsilon),$$

where $w_R(t,\varepsilon)$ is given in Theorem 3.1 and $\Gamma(\varepsilon) = (\varepsilon\gamma m^{-1})^{1/(2q+1)}$, for $\gamma \geq |u''|(2q+1)!$. Clearly we have $\alpha \leq \beta$, $\alpha(a,\varepsilon) - p_1\alpha'(a,\varepsilon) \leq A \leq \beta(a,\varepsilon) - p_1\beta'(a,\varepsilon)$, and $\alpha(b,\varepsilon) \leq B \leq \beta(b,\varepsilon)$. As regards the differential inequalities, let us suppose that u is (I_q)-stable and let us consider only β. (The verification for α follows by symmetry.) Expanding by Taylor's Theorem we see that

$$\begin{aligned} h(t,\beta) - \varepsilon\beta'' &= h(t,u) + \sum_{j=1}^{2q} \frac{1}{j!} \partial_y^j h(t,u)(\beta-u)^j \\ &\quad + \frac{1}{(2q+1)!} \partial_y^{2q+1} h(t,\eta)(\beta-u)^{2q+1} \\ &\quad - \varepsilon u'' - \varepsilon v_L'' - \varepsilon w_R'' \\ &\geq \frac{m}{(2q+1)!} \left\{v_L^{2q+1} + w_R^{2q+1}\right\} + \frac{\varepsilon\gamma}{(2q+1)!} \\ &\quad - \varepsilon|u''| - \varepsilon v_L'' - \varepsilon w_R'' \\ &\geq 0 \end{aligned}$$

by virtue of our assumptions. Here (t,η) is the appropriate intermediate point, which lies in $\mathcal{D}_0(u)$ provided ε is sufficiently small, say $0 < \varepsilon \leq \varepsilon_0$. Thus the conclusion of Theorem 3.4 follows from Theorem 2.3.

If the solution u of (R_1) is (II_n)- or $(\tilde{II}_n)$-stable, then as was the case with the Dirichlet problem, u must satisfy the additional requirements: $u'' \geq 0$ in (a,b), $u(a) - p_1u'(a) \leq A$ and $u(b) \leq B$. Similarly, if u is (III_n)- or $(I\tilde{I}I_n)$-stable, then we must require that $u'' \leq 0$ in (a,b), $u(a) - p_1u'(a) \geq A$ and $u(b) \geq B$. The precise results are contained in the following theorems.

Theorem 3.5. Assume that the reduced equation (R_1) has a (II_n)- or $(\tilde{II}_n)$-stable solution $u = u(t)$ of class $C^{(2)}([a,b])$ such that $u(a) - p_1u'(a) \leq A$, $u(b) \leq B$ and $u'' \geq 0$ in (a,b). Then there exists an $\varepsilon_0 > 0$ such that for $0 < \varepsilon \leq \varepsilon_0$ the problem (RP_1) has a solution $y = y(t,\varepsilon)$ in $[a,b]$ which satisfies

$$0 \leq y(t,\varepsilon) - u(t) \leq v_L(t,\varepsilon) + w_R(t,\varepsilon) + c\varepsilon^{1/n}.$$

Here

$$v_L(t,\varepsilon) = \sigma[1 + \tfrac{1}{2}(n-1)(\varepsilon(n+1)!/2m)^{-1/2}\sigma^{1/2(n-1)}(t-a)]^{-2/(n-1)},$$

where

$$\sigma^{n+1} = \varepsilon(n+1)!(A-u(a) + p_1u'(a))^2/2mp_1^2,$$

w_R is as given in Theorem 3.2, and c is some positive constant.

Proof: The proof of Theorem 3.5 is almost a repetition of the proof of the previous theorem, if we define for t in $[a,b]$ and $\varepsilon > 0$ the functions

$$\alpha(t,\varepsilon) = u(t)$$

and

$$\beta(t,\varepsilon) = u(t) + v_L(t,\varepsilon) + w_R(t,\varepsilon) + (\varepsilon\gamma m^{-1})^{1/n},$$

where $\gamma \geq |u''|n!$. The details are left to the reader.

Theorem 3.6. Assume that the reduced equation (R_1) has a (III_n)- or $(\tilde{III}_n)$-stable solution $u = u(t)$ of class $C^{(2)}([a,b])$ such that $u(a) - p_1u'(a) \geq A$, $u(b) \geq B$ and $u'' \leq 0$ in (a,b). Then there exists an $\varepsilon_0 > 0$ such that for $0 < \varepsilon \leq \varepsilon_0$ the problem (RP_1) has a solution $y = y(t,\varepsilon)$ for t in $[a,b]$ which satisfies

$$-v_L(t,\varepsilon) - w_R(t,\varepsilon) - c\varepsilon^{1/n} \leq y(t,\varepsilon) - u(t) \leq 0,$$

where v_L and w_R are as defined in the conclusion of Theorem 3.5.

The proof follows if we simply let $y \to -y$ and apply Theorem 3.5 to the transformed problem.

It is now an easy matter to discuss the behavior of solutions of the problem (RP_2). For this problem, a solution $u = u(t)$ of the reduced equation (R_1) is said to be (I_q)-, (II_n)- or (III_n)- stable if it is so stable in the sense of Definitions 3.1, 3.3 or 3.5, respectively, with $d(t) \equiv \delta$ in $[a,b]$. The proofs of the next two results can be patterned after those of Theorems 3.4 and 3.5 and are omitted.

Theorem 3.7. Assume that the reduced equation (R_1) has an (I_q)-stable solution $u = u(t)$ of class $C^{(2)}([a,b])$. Then there exists an $\varepsilon_0 > 0$ such that for $0 < \varepsilon \leq \varepsilon_0$ the problem (RP_2) has a solution $y = y(t,\varepsilon)$ in $[a,b]$ which satisfies

$$|y(t,\varepsilon)-u(t)| \leq v_L(t,\varepsilon) + v_R(t,\varepsilon) + c\varepsilon^{1/(2q+1)}.$$

Here v_L is as given in the conclusion of Theorem 3.4 and

$$v_R(t,\varepsilon) = (\varepsilon/mp_2^2)^{1/2}|B-u(b)-p_2u'(b)| \exp[-(m/\varepsilon)^{1/2}(b-t)] \quad \text{if } q = 0,$$

$$v_R(t,\varepsilon) = \sigma[1 + q(\varepsilon(2q+2)!/2m)^{-1/2}\sigma^q(b-t)]^{-1/q} \quad \text{if } q \geq 1,$$

where $\sigma^{q+1} = \{\varepsilon(2q+2)!/2mp_2^2\}^{1/2}|B-u(b)-p_2u'(b)|$ and c is some positive constant.

Theorem 3.8. Assume that the reduced equation (R_1) has a (II_n)-stable solution $u = u(t)$ of class $C^{(2)}([a,b])$ such that $u(a) - p_1u'(a) \leq A$, $u(b) + p_2u'(b) \leq B$ and $u'' \geq 0$ in (a,b). Then there exists an $\varepsilon_0 > 0$ such that for $0 < \varepsilon \leq \varepsilon_0$ the problem (RP_2) has a solution $y = y(t,\varepsilon)$ in $[a,b]$ which satisfies

$$0 \leq y(t,\varepsilon) - u(t) \leq v_L(t,\varepsilon) + v_R(t,\varepsilon) + c\varepsilon^{1/n}.$$

Here v_L is as given in the conclusion of Theorem 3.5, and

$$v_R(t,\varepsilon) = \sigma[1 + \tfrac{1}{2}(n-1)(\varepsilon(n+1)!/2m)^{-1/2}\sigma^{1/2(n-1)}(b-t)]^{-2/(n-1)},$$

where $\sigma^{n+1} = \varepsilon(n+1)!(B-u(b)-p_2u'(b))^2/2mp_2^2$ and c is some positive constant.

The corresponding result for the problem (RP_2) in the case that u is (III_n)-stable and satisfies $u(a) - p_1u'(a) \geq A$, $u(b) + p_2u'(b) \geq B$ and $u'' \leq 0$ in (a,b) follows from Theorem 3.8 after making the change of variable $y \to -y$. We leave its precise formulation to the reader.

§3.3. Interior Layer Phenomena

The above results deal with stable solutions $u = u(t)$ of $h(t,u) = 0$ which are twice continuously differentiable in $[a,b]$. The smoothness restriction imposed on u can be slightly weakened without altering the validity of these results. For example, it is enough to assume that the second derivative of u is bounded in $[a,b]$ or even that the function u' is differentiable almost everywhere in (a,b), with u'' bounded wherever it exists. Let us now suppose that the continuous function u is of class $C^{(2)}([a,b])$ except at the point t_0 in (a,b) where $u'(t_0^-) \neq u'(t_0^+)$. It is easy to see that such a situation can arise in the type of problems we have been considering. Namely, if two $C^{(2)}$-solutions u_1 and u_2 of $h(t,u) = 0$ intersect at the point t_0 in (a,b) with $u_1'(t_0) \neq u_2'(t_0)$, then the path $u_0(t)$ defined by

$$u_0(t) = \begin{cases} u_1(t), & a \leq t \leq t_0, \\ u_2(t), & t_0 \leq t \leq b, \end{cases}$$

will have the property that $u_0'(t_0^-) \neq u_0'(t_0^+)$. If both functions u_1 and u_2 are stable in $[a,b]$, then the path u_0 is also stable, and it is reasonable to expect that, under appropriate restrictions on $u_0'(t_0^{\pm})$, there is a solution $y = y(t,\varepsilon)$ of the problem (DP_1), (RP_1) or (RP_2) such that

$$\lim_{\varepsilon\to 0^+} y(t,\varepsilon) = u_0(t) \quad \text{in each closed subinterval of } (a,b).$$

This will turn out to be the case if we supplement the bounding functions α,β with an "interior layer corrector at t_0". For ease of exposition, we assume that the reduced path u is not differentiable at only one single point t_0 in (a,b), and that u is either q- or n-stable throughout $[a,b]$. The extension of our results to the case of finitely many points of nondifferentiability and to the case when u is, say, q-stable in (a,t_0) and n-stable in (t_0,b) is rather straightforward and will be omitted.

Theorem 3.9. Assume that the reduced equation (R_1) has an (I_q)-or $(\tilde{I}_q)$-stable solution $u = u(t)$ of class $C^{(2)}([a,b])$, except at t_0 in (a,b) where $u'(t_0^-) \neq u'(t_0^+)$ and $|u''(t_0^{\pm})| < \infty$. Then there exists an $\varepsilon_0 > 0$ such that for $0 < \varepsilon \leq \varepsilon_0$ the problem (DP_1), (RP_1) or (RP_2) has a solution $y = y(t,\varepsilon)$ for t in $[a,b]$ which satisfies, respectively,

$$|y(t,\varepsilon)-u(t)| \le w_L(t,\varepsilon) + w_R(t,\varepsilon) + v_I(t,\varepsilon) + c\varepsilon^{1/(2q+1)} \qquad (DP_1)$$

$$|y(t,\varepsilon)-u(t)| \le v_L(t,\varepsilon) + w_R(t,\varepsilon) + v_I(t,\varepsilon) + c\varepsilon^{1/(2q+1)} \qquad (RP_1)$$

and

$$|y(t,\varepsilon)-u(t)| \le v_L(t,\varepsilon) + v_R(t,\varepsilon) + v_I(t,\varepsilon) + c\varepsilon^{1/(2q+1)} \qquad (RP_2).$$

Here w_L and w_R are as given in Theorem 3.1, v_L and v_R are as given in Theorem 3.7, and

$$v_I(t,\varepsilon) = \frac{1}{2}(m^{-1}\varepsilon)^{1/2}|u'(t_0+)-u'(t_0-)|\exp[-(m\varepsilon^{-1})^{1/2}|t-t_0|] \quad \text{if } q = 0,$$

$$v_I(t,\varepsilon) = \frac{1}{2}\,\sigma[1 + q(\varepsilon(2q+2)!/2m)^{-1/2}\sigma^q|t-t_0|]^{-1/q} \quad \text{if } q \ge 1,$$

where $\sigma^{q+1} = |u'(t_0+)-u'(t_0-)|\{\varepsilon(2q+2)!/2m\}^{1/2}$ and c is some positive constant.

Proof: Let us first consider the Dirichlet problem (DP_1). We can suppose that $u'(t_0^-) < u'(t_0^+)$. (If $u'(t_0^-) > u'(t_0^+)$, we let $y \to -y$ and obtain this case.) Then we define for t in $[a,b]$ and $\varepsilon > 0$

$$\alpha(t,\varepsilon) = u(t) - \Gamma(\varepsilon),$$

$$\beta(t,\varepsilon) = u(t) + w_L(t,\varepsilon) + w_R(t,\varepsilon) + v_I(t,\varepsilon) + \Gamma(\varepsilon),$$

where $\Gamma(\varepsilon) = (\varepsilon\gamma m^{-1})^{1/(2q+1)}$ for $\gamma \ge |u''|(2q+1)!$. The function α is not differentiable at $t = t_0$; however, this presents no problem because $\alpha'(t_0^-) < \alpha'(t_0^+)$. Indeed, for t in $(a,t_0) \cup (t_0,b)$, we have $\varepsilon\alpha'' \ge h(t,\alpha)$, and so α is a lower solution there (cf. Theorem 2.2). For the function β, we note that v_I is the solution of $\varepsilon v'' = \frac{m}{(2q+1)!}\,v^{2q+1}$ in $(a,t_0) \cup (t_0,b)$ which satisfies

$$v_I'(t_0^-,\varepsilon) = -v_I'(t_0^+,\varepsilon) = \frac{1}{2}|u'(t_0^+)-u'(t_0^-)|,$$

$$v_I(t_0^-,\varepsilon) = v_I(t_0^+,\varepsilon) = \sigma.$$

With this function v_I, we see that β is differentiable at $t = t_0$; indeed,

$$\beta'(t_0^-,\varepsilon) = \beta'(t_0^+,\varepsilon) = \frac{1}{2}|u'(t_0^-)+u'(t_0^+)| + w_L'(t_0,\varepsilon) + w_R'(t_0,\varepsilon)$$

and, as before, we can show that $\varepsilon\beta'' \le h(t,\beta)$. Thus (α,β) is a bounding pair and the result for the problem (DP_1) follows from Theorem 2.2.

Similarly we can treat the problems (RP_1) and (RP_2); we leave the details to the reader.

If the reduced path u is (II_n)- or $(\tilde{II}_n)$-stable, then the requirement of convexity (that is, $u'' \geq 0$) must be interpreted as follows:

$$u'' \geq 0 \quad \text{in} \quad (a,t_0) \cup (t_0,b) \quad \text{and} \quad u'(t_0^-) < u'(t_0^+).$$

The precise result is contained in the next theorem whose proof is similar to that of Theorem 3.9.

Theorem 3.10. Assume that the reduced equation (R_1) has a (II_n)-or $(\tilde{II}_n)$-stable solution $u = u(t)$ of class $C^{(2)}([a,b])$, except at t_0 in (a,b) where $u'(t_0^-) < u'(t_0^+)$ and $|u''(t_0^\pm)| < \infty$. Assume also that $u(a) - \pi_1 u'(a) \leq A$ (for $\pi_1 = p_1$ or 0), $u(b) + \pi_2 u'(b) \leq B$ (for $\pi_2 = p_2$ or 0) and $u'' \geq 0$ in $(a,t_0) \cup (t_0,b)$. Then there exists an $\varepsilon_0 > 0$ such that for $0 < \varepsilon \leq \varepsilon_0$ the problem (DP_1), (RP_1) or (RP_2) has a solution $y = y(t,\varepsilon)$ for t in $[a,b]$ which satisfies, respectively,

$$0 \leq y(t,\varepsilon) - u(t) \leq w_L(t,\varepsilon) + w_R(t,\varepsilon) + v_I(t,\varepsilon) + c\varepsilon^{1/n} \qquad (DP_1),$$

$$0 \leq y(t,\varepsilon) - u(t) \leq v_L(t,\varepsilon) + w_R(t,\varepsilon) + v_I(t,\varepsilon) + c\varepsilon^{1/n} \qquad (RP_1)$$

or

$$0 \leq y(t,\varepsilon) - u(t) \leq v_L(t,\varepsilon) + v_R(t,\varepsilon) + v_I(t,\varepsilon) + c\varepsilon^{1/n} \qquad (RP_2).$$

Here w_L and w_R are as given in Theorem 3.2, v_L and v_R are as given in Theorem 3.8, and

$$v_I(t,\varepsilon) = \tfrac{1}{2}\sigma[1 + \tfrac{1}{2}(n-1)(\varepsilon(n+1)!/2m)^{-1/2}\sigma^{1/2(n-1)}|t-t_0|]^{-2/(n-1)},$$

where

$$\sigma^{n+1} = \varepsilon(n+1)!|u'(t_0^+) - u'(t_0^-)|^2/2m$$

and c is some positive constant.

Finally, if the reduced path u is (III_n)- or $(\tilde{III}_n)$-stable, then the result analogous to Theorem 3.10 is valid provided that $u(a) - \pi_1 u'(a) \geq A$, $u(b) + \pi_2 u'(b) \geq B$, $u'' \leq 0$ in $(a,t_0) \cup (t_0,b)$ and $u'(t_0^-) > u'(t_0^+)$.

Notes and Remarks

3.1. The theory of this chapter applies with little change to the more general problem $\varepsilon y'' = h(t,y,\varepsilon)$, $a < t < b$, $y(a,\varepsilon) = A(\varepsilon)$, $y(b,\varepsilon) = B(\varepsilon)$. We need only require that $h(t,y,\varepsilon) = h(t,y,0) + o(1)$ for (t,y) in $\mathcal{D}_i(u)$ $(i = 0,1,2)$ and that $A(\varepsilon) = A(0) + o(1)$ and $B(\varepsilon) = B(0) + o(1)$ for all sufficiently small values of ε.

3.2. The definitions of (I_q)-, (II_n)- and (III_n)-stability were introduced by Boglaev [6] and used by him to study the Dirichlet problem (DP_1). Earlier Bris proved Theorem 3.1 in the case of (I_0)-stability. Among the other work done on the problem (DP_1) we mention only the papers of Tupchiev [86], Vasil'eva [87], Vasil'eva and Tupchiev [89], Carrier [8], Fife [24], O'Malley [76], Dorr, Parter and Shampine [20], Habets [29], Habets and Laloy [31], Flaherty and O'Malley [25] and Howes [39]. The Robin problems (RP_1) and (RP_2) have also been considered by Habets and Laloy in the case of (I_0)-stability.

3.3. The stability requirements on the solution u of the reduced equation can be relaxed as follows (cf. Fife [24], Flaherty and O'Malley [25] and Howes [39]). Namely it is enough in the case of (I_q)- or $(\tilde{I}_q)$-stability that

$$\partial_y^j h(t,u(t)) \equiv 0 \quad \text{for} \quad 0 \le j \le 2q,$$

$$\partial_y^{2q+1} h(t,u(t)) \ge m > 0 \quad \text{in} \quad [a,b],$$

and for $u(a) \neq A$,

$$\int_{u(a)}^{\xi} h(a,s)ds > 0 \quad \text{for} \quad \xi \quad \text{in} \quad [A,u(a)) \quad \text{or} \quad (u(a),A]$$

or, for $u(b) \neq B$,

$$\int_{u(b)}^{\eta} h(b,s)ds > 0 \quad \text{for} \quad \eta \quad \text{in} \quad [B,u(b)) \quad \text{or} \quad (u(b),B].$$

Similar relaxations apply to the cases of (II_n)-stability $(u(a) \le A,\ u(b) \le B)$ and (III_n)-stability $(u(a) \ge A,\ u(b) \ge B)$.

3.4. We have not considered the occurrence of shock layer behavior, that is, the situation in which a solution $y = y(t,\varepsilon)$ of (DP_1), (RP_1) or (RP_2) satisfies the limiting relation

$$\lim_{\varepsilon \to 0^+} y(t,\varepsilon) = \begin{cases} u_1(t), & a < t < t_0, \\ u_2(t), & t_0 < t < b, \end{cases}$$

where $u_1(t_0) \neq u_2(t_0)$. The functions u_1 and u_2 are stable solutions of the reduced equation (R_1). These phenomena are studied, for instance, by Vasil'eva [88], Fife [24], O'Malley [76] and Howes [39] to which the reader can refer for details.

3.5. Oscillatory phenomena of the type exhibited by the solution of the problem (E_2) are discussed for more general problems by Volosov [91] and O'Malley [76].

3.6. We note that in the case of the Robin problem (RP_2), (I_q)-, (II_n)- and (III_n)-stability are essentially equivalent to their "tilded" counterparts because $\mathcal{D}_i(u)$ $(i = 0,1,2)$ is a "δ-tube" around the function $u(t)$, and δ can be taken arbitrarily small.

3.7. The theory developed in this chapter for the Robin problems (RP_1) and (RP_2) applies with minor modification to the Neumann problem $\varepsilon y'' = h(t,y)$, $a < t < b$, $-y'(a,\varepsilon) = A$, $y'(b,\varepsilon) = B$, and related problems.

Chapter IV
Quasilinear Singular Perturbation Problems

§4.1. The Dirichlet Problem: Boundary Layer Phenomena

We consider now the singularly perturbed quasilinear Dirichlet problem

$$\varepsilon y'' = f(t,y)y' + g(t,y) \equiv F(t,y,y'), \quad a < t < b,$$
$$y(a,\varepsilon) = A, \quad y(b,\varepsilon) = B. \tag{DP_2}$$

If $f(t,y) \not\equiv 0$, a great variety of interesting phenomena can occur. If $f(t,y) \equiv 0$, the problem (DP_2) is identical to the problem (DP_1) already discussed in the previous chapter. Therefore at points (t,y) for which $f(t,y) = 0$, we require the function $F(t,y,y')$ to be stable with respect to the y-variable (in the sense of Definitions 3.1-3.6), in order that the solutions of (DP_2) may behave "reasonably". There are, however, qualitative differences between the two problems (DP_1) and (DP_2) which will be illustrated by the following simple examples.

Consider first the linear problems

$$\varepsilon y'' = \pm y', \quad 0 < t < 1,$$
$$y(0,\varepsilon) = 0, \quad y(1,\varepsilon) = 1. \tag{$E_6^\pm$}$$

The solution of (E_6^+) is easily found to be

$$y(t,\varepsilon) = (1-e^{-1/\varepsilon})^{-1}[-e^{-1/\varepsilon} + e^{-(1-t)/\varepsilon}] \sim e^{-(1-t)/\varepsilon},$$

and so

$$\lim_{\varepsilon\to 0^+} y(t,\varepsilon) = 0 \quad \text{for} \quad 0 \le t \le 1 - \delta < 1, \tag{4.1}$$

where δ, $0 < \delta < 1$, is a fixed constant. That is, the solution only exhibits nonuniform convergence at the right endpoint $t = 1$. The limiting function $u \equiv 0$ is the solution of the corresponding reduced equation $u' = 0$ which satisfies $u(0) = 0 = y(0,\varepsilon)$.

On the other hand, the problem (E_6^-) has the solution

$$y(t,\varepsilon) = (e^{-1/\varepsilon}-1)^{-1}(-1+e^{-t/\varepsilon}) \sim 1 - e^{-t/\varepsilon}$$

and so

$$\lim_{\varepsilon\to 0^+} y(t,\varepsilon) = 1 \quad \text{for} \quad 0 < \delta \le t \le 1. \tag{4.2}$$

That is, the solution exhibits nonuniform convergence at the left endpoint $t = 0$. Here $u \equiv 1$ is a solution of the reduced equation $u' = 0$ satisfying $u(1) = 1 = y(1,\varepsilon)$. Thus the nature of the solutions of these two problems depends critically on the sign of the coefficient of y'. It is interesting to note also that the relations (4.1) and (4.2) differ from (3.2) in that nonuniform convergence of the solution (that is, boundary layer behavior) occurs at *one* endpoint only.

Consider next the problem

$$\begin{aligned} &\varepsilon y'' = -ty', \quad -1 < t < 1, \\ &y(-1,\varepsilon) = -1, \quad y(1,\varepsilon) = 1. \end{aligned} \tag{E_7}$$

The exact solution is

$$y(t,\varepsilon) = -1 + 2\left(\int_{-1}^{1} e^{-s^2/2\varepsilon}ds\right)^{-1} \int_{-1}^{t} e^{-s^2/2\varepsilon}ds,$$

from which it follows that

$$\lim_{\varepsilon\to 0^+} y(t,\varepsilon) = \begin{cases} -1 & \text{for} \quad -1 \le t \le -\delta < 0, \\ 1 & \text{for} \quad 0 < \delta \le t \le 1, \end{cases}$$

where δ, $0 < \delta < 1$, is a fixed constant. The function $u_1 \equiv -1$ is the solution of the reduced equation $tu' = 0$ satisfying $u_1(-1) = -1 = y(-1,\varepsilon)$, while $u_2 \equiv 1$ is the solution of $tu' = 0$ satisfying $u_2(1) = 1 = y(1,\varepsilon)$. In this case, the solution exhibits uniform convergence at both endpoints. This is not at all surprising, if we note that near the left endpoint $t = -1$, the coefficients of y' in (E_7) and in (E_6^+) are both positive, and therefore we do not anticipate nonuniform convergence of the solution at the left endpoint $t = -1$. Similarly, near the right endpoint $t = 1$, the coefficients of y' in (E_7) and in (E_6^-) are both negative, and so we do not anticipate nonuniform convergence of the solution at the right endpoint $t = 1$. Instead, the solution exhibits nonuniform convergence at an interior point $t = 0$, where it switches (discontinuously in the limit as $\varepsilon \to 0$) from $u_1 \equiv -1$ to $u_2 \equiv 1$.

As our third example we take the problem

$$\begin{aligned} &\varepsilon y'' = ty' + y, \quad -1 < t < 1, \\ &y(-1,\varepsilon) = 1, \quad y(1,\varepsilon) = 2. \end{aligned} \tag{E_8}$$

The solution $y = y(t,\varepsilon)$ of (E_8) can be shown to satisfy

$$\lim_{\varepsilon\to 0^+} y(t,\varepsilon) = 0 \quad \text{for} \quad -1 < -1 + \delta \le t \le 1 - \delta < 1, \tag{4.3}$$

where δ, $0 < \delta < 1$, is a fixed constant. The coefficient of y' is negative (positive) near $t = -1$ $(t = 1)$, and so the nonuniform convergence of the solution near both endpoints is in accordance with that observed in $(E_6^{\pm})$. The limiting function $u \equiv 0$ is the continuous solution of the reduced equation $tu' + u = 0$, whose general solution $u = c/t$ is not continuous at $t = 0$.

Our final example is

$$\begin{aligned} &\varepsilon y'' = yy', \quad -1 < t < 1, \\ &y(-1,\varepsilon) = A, \quad y(1,\varepsilon) = B, \end{aligned} \tag{E_9}$$

which has been discussed in greater detail by Wasow [94] and O'Malley [75; Chapter 1]. The algebraic sign of the coefficient of y' in this equation depends on the solution under consideration, and so the behavior exhibited by solutions of (E_9) is more varied than that previously encountered. For example, if $A > 0$ and $A > -B$, then the solution $y = y(t,\varepsilon)$ of (E_9) satisfies

$$\lim_{\varepsilon\to 0+} y(t,\varepsilon) = A \quad \text{for} \quad -1 \le t \le 1 - \delta < 1$$

(cf. (E_6^+)), while if $B < 0$ and $A < -B$, then

$$\lim_{\varepsilon\to 0+} y(t,\varepsilon) = B \quad \text{for} \quad -1 < -1 + \delta \le t \le 1$$

(cf. (E_6^-)). On the other hand, if $A = -B > 0$ then the solution $y = y(t,\varepsilon)$ of (E_9) satisfies

$$\lim_{\varepsilon\to 0+} y(t,\varepsilon) = \begin{cases} A & \text{for} \quad -1 \le t \le -\delta < 0, \\ B & \text{for} \quad 0 < \delta \le t \le 1, \end{cases}$$

(cf. (E_7)), while if $A < 0 < B$, then

$$\lim_{\varepsilon\to 0+} y(t,\varepsilon) = 0 \quad \text{for} \quad -1 < -1 + \delta \le t \le 1 - \delta < 1$$

(cf. (E_8)).

As these examples show, the phenomena exhibited by solutions of the problem (DP_2) are more diverse and complicated than those encountered in the last chapter. Our purpose here is to isolate several classes of problems for which coherent results are possible and to refer the reader to the relevant literature for discussions of current problems where complete answers are as yet unknown.

We begin our study of the problem (DP_2) by associating with it three types of reduced problems, namely,

$$f(t,u)u' + g(t,u) = 0, \quad a < t < t_1 \le b, \qquad u(a) = A, \tag{R_L}$$

$$f(t,u)u' + g(t,u) = 0, \quad a \le t_2 < t < b, \qquad u(b) = B, \tag{R_R}$$

and

$$f(t,u)u' + g(t,u) = 0, \quad a < t < b, \tag{R}$$

where t_1, t_2 are fixed numbers. In what follows, solutions of (R_L), (R_R) and (R) will be denoted by u_L, u_R and u, respectively. Note that the function u_L is only required to exist in $[a,t_1) \subset [a,b]$, while u_R is only defined in $(t_2,b] \subset [a,b]$ for $a \le t_1$, $t_2 \le b$.

As in the previous chapter, we will only study solutions of these reduced problems which are stable in a sense to be defined momentarily. The definitions of stability will be given in terms of the following domains, where $\delta > 0$ is a small constant.

If a solution $u = u_L(t)$ of (R_L) exists in $[a,b]$, then we define the domain $\mathcal{D}(u_L)$ by

$$\mathcal{D}(u_L) = \{(t,y) : a \le t \le b, \ |y-u_L(t)| \le d_L(t)\},$$

where the positive continuous function $d_L(t)$ satisfies $d_L(t) \equiv |B-u_L(b)| + \delta$ for $b - \delta/2 \le t \le b$ and $d_L(t) \equiv \delta$ for $a \le t \le b - \delta$.

Similarly, if a solution $u = u_R(t)$ of (R_R) exists in $[a,b]$ then we define the domain $\mathcal{D}(u_R)$ by

$$\mathcal{D}(u_R) = \{(t,y): a \le t \le b, \ |y-u_R(t)| \le d_R(t)\}$$

where the positive continuous function $d_R(t)$ satisfies $d_R(t) \equiv |A-u_R(a)| + \delta$ for $a \le t \le a + \delta/2$ and $d_R(t) \equiv \delta$ for $a+\delta \le t \le b$.

Finally, for a solution $u = u(t)$ of (R) we define the domain $\mathcal{D}(u)$ by

$$\mathcal{D}(u) = \{(t,y)\colon a \le t \le b,\ |y-u(t)| \le d(t)\},$$

where the positive continuous function $d(t)$ satisfies $d(t) \equiv |A-u(a)| + \delta$ for $a \le t \le a + \delta/2$, $d(t) \equiv |B-u(b)| + \delta$ for $b - \delta/2 \le t \le b$ and $d(t) \equiv \delta$ for $a + \delta \le t \le b - \delta$. In addition, we will also consider paths of the form

$$u_0(t) = \begin{cases} u_L(t), & a \le t \le t_L \le t_1, \\ u(t), & t_L \le t \le t_R \le t_2, \\ u_R(t), & t_R \le t \le b, \end{cases}$$

$$u_1(t) = \begin{cases} u(t), & a \le t \le t_R, \\ u_R(t), & t_R \le t \le b, \end{cases}$$

and

$$u_2(t) = \begin{cases} u_L(t), & a \le t \le t_L, \\ u(t), & t_L \le t \le b, \end{cases}$$

for $a < t_L \le t_R < b$, and we define the additional domains:

$$\mathcal{D}(u_L,u,u_R) = \{(t,y)\colon a \le t \le b,\ |y-u_0(t)| \le \delta\},$$

$$\mathcal{D}(u_L,u_R) = \mathcal{D}(u_L,u,u_R) \quad \text{with} \quad t_L = t_R,$$

$$\mathcal{D}(u,u_R) = \{(t,y)\colon a \le t \le b,\ |y-u_1(t)| \le d_1(t)\},$$

where the positive continuous function d_1 satisfies $d_1(t) \equiv |A-u(a)| + \delta$ for $a \le t \le a + \delta/2$ and $d_1(t) \equiv \delta$ for $a + \delta \le t \le b$, and

$$\mathcal{D}(u_L,u) = \{(t,y)\colon a \le t \le b,\ |y-u_2(t)| \le d_2(t)\}$$

where the positive continuous function d_2 satisfies $d_2(t) \equiv |B-u(b)| + \delta$ for $b - \delta/2 \le t \le b$ and $d_2(t) \equiv \delta$ for $a \le t \le b - \delta$.

We can now define the types of stability which we will require of solutions of the reduced problems (R_L), (R_R) and (R).

<u>Definition 4.1</u>. A solution $u = u_L(t)$ of the reduced problem (R_L) is said to be strongly (weakly) stable in $[a,b]$ if there exists a positive constant k such that

$$f(t,y) \ge k > 0 \quad (f(t,y) \ge 0) \quad \text{in} \quad \mathcal{D}(u_L).$$

Definition 4.2. A solution $u = u_R(t)$ of the reduced problem (R_R) is said to be strongly (weakly) stable in $[a,b]$ if there exists a positive constant k such that

$$f(t,y) \leq -k < 0 \quad (f(t,y) \leq 0) \quad \text{in } \mathcal{D}(u_R).$$

Definition 4.3. A solution $u = u(t)$ of the reduced equation (R) is said to be locally strongly (weakly) stable if there exists a positive constant k and a small positive constant δ such that, if $u(a) \neq A$,

$$f(t,y) \leq -k < 0 \quad (f(t,y) \leq 0) \quad \text{in } \mathcal{D}(u) \cap [a, a + \tfrac{\delta}{2}]$$

and, if $u(b) \neq B$,

$$f(t,y) \geq k > 0 \quad (f(t,y) \geq 0) \quad \text{in } \mathcal{D}(u) \cap [b - \tfrac{\delta}{2}, b].$$

Definition 4.4. A solution $u = u(t)$ of the reduced equation (R) is said to be (I_q)-, $(\tilde{I}_q)$-, (II_n)-, $(\tilde{II}_n)$-, (III_n)- or $(\tilde{III}_n)$-stable in $[a,b]$ if the function $h(t,y) = f(t,y)u'(t) + g(t,y)$ satisfies all the requirements in the Definitions 3.1 - 3.6 for (I_q)-, ..., $(\tilde{III}_n)$-stability, respectively.

Definition 4.5. The reduced path $u = u_0(t)$ is said to be weakly stable in $[a,b]$ if there exists a small positive constant δ such that

$$f(t,y) \geq 0 \quad \text{in } \mathcal{D}(u_L,u,u_R) \cap ([a,t_L] \cup [t_R - \tfrac{\delta}{2}, t_R])$$

and

$$f(t,y) \leq 0 \quad \text{in } \mathcal{D}(u_L,u,u_R) \cap ([t_L, t_L + \tfrac{\delta}{2}] \cup [t_R,b]).$$

If $t_L = t_R$ $(= t_0)$, then $\mathcal{D}(u_L,u,u_R) = \mathcal{D}(u_L,u_R)$; and the reduced path

$$u_0(t) = \begin{cases} u_L(t), & a \leq t \leq t_0, \\ u_R(t), & t_0 \leq t \leq b, \end{cases}$$

is said to be weakly stable in $[a,b]$ if

$$f(t,y) \geq 0 \quad \text{in } \mathcal{D}(u_L,u_R) \cap [a,t_0]$$

and

$$f(t,y) \leq 0 \quad \text{in } \mathcal{D}(u_L,u_R) \cap [t_0,b].$$

Definition 4.6. The reduced path $u = u_1(t)$ is said to be weakly stable in $[a,b]$ if there exists a small positive constant δ such that

$$f(t,y) \geq 0 \quad \text{in} \quad \mathcal{D}(u,u_R) \cap [t_R - \tfrac{\delta}{2}, t_R]$$

and

$$f(t,y) \leq 0 \quad \text{in} \quad \mathcal{D}(u_L,u) \cap [t_L, t_L + \tfrac{\delta}{2}].$$

Definition 4.7. The reduced path $u = u_2(t)$ is said to be weakly stable in $[a,b]$ if there exists a small positive constant δ such that

$$f(t,y) \geq 0 \quad \text{in} \quad \mathcal{D}(u_L,u) \cap [a,t_L]$$

and

$$f(t,y) \leq 0 \quad \text{in} \quad \mathcal{D}(u_L,u) \cap [t_L, t_L + \tfrac{\delta}{2}].$$

With these definitions we can now discuss the asymptotic behavior of solutions of the Dirichlet problem (DP_2). We assume throughout this chapter that the functions f and g are continuous in t and y and of class $C^{(1)}$ with respect to y in the domain under consideration. Moreover, in the cases of q- or n-stable reduced solutions, we assume that f and g have the required number of continuous partial derivatives with respect to y.

Finally, in the conclusion of each theorem below, the constant c is a known positive constant depending on the reduced solution under consideration.

Theorem 4.1. Let the reduced problem (R_R) have a strongly stable solution $u = u_R(t)$ of class $C^{(2)}([a,b])$. Then there exists an $\varepsilon_0 > 0$ such that for $0 < \varepsilon \leq \varepsilon_0$ the problem (DP_2) has a solution $y = y(t,\varepsilon)$ for t in $[a,b]$ which satisfies

$$|y(t,\varepsilon) - u_R(t)| \leq w_L(t,\varepsilon) + c\varepsilon,$$

where

$$w_L(t,\varepsilon) = |A - u_R(a)| \exp[\lambda(t-a)]$$

and $\lambda = -k/\varepsilon + \ell/k + O(\varepsilon)$.

Proof: As in the proof of Theorem 3.1, we construct a bounding pair of functions (α,β) which satisfy all the conditions of Theorem 2.1.

Let us first linearize (DP_2) about the reduced solution u_R by setting $z = y - u_R$. This leads to

$$\varepsilon z'' = f(t,y)z' + h_y(t,\xi)z - \varepsilon u_R'',$$

$$z(a,\varepsilon) = A - u_R(a), \quad z(b,\varepsilon) = 0,$$

where $h_y(t,\xi) = f_y(t,\xi)u_R' + g_y(t,\xi)$ and (t,ξ) is some intermediate point between (t,u) and $(t, u+z)$. Since $f(t,y) \leq -k < 0$ by definition of strong stability, and since h_y is bounded, say $|h_y| \leq \ell$ $(\ell > 0)$, in $\mathcal{D}(u_R)$, we are further led to the linear non-homogeneous differential equation

$$\varepsilon z'' + kz' + \ell z = -\varepsilon u_R''.$$

If $0 < \varepsilon < k^2/4\ell$, the corresponding auxiliary equation $\varepsilon\lambda^2 + k\lambda + \ell = 0$ has two negative roots

$$\lambda = -k/\varepsilon + \ell/k + O(\varepsilon) = -k/\varepsilon + O(1)$$

$$\lambda_1 = -\ell/k + O(\varepsilon) = O(1).$$

Indeed, the positive function $w_L(t,\varepsilon) = |A - u_R(a)|\exp[\lambda(t-a)]$ is the solution of the corresponding homogeneous differential equation

$$\varepsilon w'' + kw' + \ell w = 0$$

such that

$$w_L(a,\varepsilon) = |A - u_R(a)|, \quad w_L'(a,\varepsilon) = \lambda|A - u_R(a)| < 0,$$

that is, the solution w_L decreases to the right.

Let $\gamma \geq |u_R''|$. Then the function $\Gamma(t,\varepsilon) = \varepsilon\gamma\ell^{-1}(\exp[-\lambda_1(b-t)] - 1)$ is such that $0 \leq \Gamma \leq c\varepsilon$ and $-c\varepsilon \leq \Gamma' < 0$ for some $c > 0$, and it satisfies

$$\varepsilon\Gamma'' + k\Gamma' + \ell\Gamma = -\varepsilon\gamma, \quad \Gamma(b) = 0.$$

We now define the bounding pair

$$\alpha(t,\varepsilon) = u_R(t) - w_L(t,\varepsilon) - \Gamma(t,\varepsilon),$$

$$\beta(t,\varepsilon) = u_R(t) + w_L(t,\varepsilon) + \Gamma(t,\varepsilon),$$

and proceed to show that they satisfy all the required inequalities of Theorem 2.1. Clearly, $\alpha \leq \beta$, $\alpha(a,\varepsilon) \leq A \leq \beta(a,\varepsilon)$ and $\alpha(b,\varepsilon) \leq B \leq \beta(b,\varepsilon)$. As regards the differential inequalities, we will only verify for α, since the verification for β follows by symmetry. By differentiation and Taylor's Theorem (with (t,ξ) as the intermediate point), we obtain

$$\begin{aligned}\varepsilon\alpha'' - F(t,\alpha,\alpha') &= \varepsilon u_R'' - \varepsilon w_L'' - \varepsilon\Gamma'' - f(t,\alpha)(\alpha'-u_R') - h_y(t,\xi)(\alpha-u_R)\\ &\geq -\varepsilon|u_R''| - \varepsilon w_L'' - \varepsilon\Gamma'' - kw_L' - k\Gamma' - \ell w_L - \ell\Gamma\\ &\geq -\varepsilon|u_R''| + \varepsilon\gamma\\ &\geq 0,\end{aligned}$$

since $\gamma \geq |u_R''|$.

Thus, α is a lower solution, and similarly β is an upper solution, and we conclude from Theorem 2.1 that for $0 < \varepsilon < k^2(4\ell)^{-1}$ the problem (DP_2) has a solution $y = y(t,\varepsilon)$ in $[a,b]$ satisfying

$$\alpha(t,\varepsilon) \leq y(t,\varepsilon) \leq \beta(t,\varepsilon).$$

In the same way we can prove the following result.

Corollary 4.1. Let the reduced problem (R_L) have a strongly stable solution $u = u_L(t)$ of class $C^{(2)}([a,b])$. Then there exists an $\varepsilon_0 > 0$ such that for $0 < \varepsilon \leq \varepsilon_0$ the problem (DP_2) has a solution $y = y(t,\varepsilon)$ for t in $[a,b]$ which satisfies

$$|y(t,\varepsilon) - u_L(t)| \leq |B - u_L(b)| \exp [\lambda(t-b)] + c\varepsilon,$$

where $\lambda = k/\varepsilon - \ell/k + O(\varepsilon) = k/\varepsilon + O(1)$.

If the functions u_L and u_R are only weakly or locally weakly stable in $[a,b]$, then we must in addition assume that u_L and u_R are y-stable in order to prove the existence of solutions of (DP_2) behaving as in Theorem 4.1. The precise results follow.

Theorem 4.2. Let the reduced problem (R_L) or (R_R) have a weakly or locally weakly stable solution $u = u_L(t)$ or $u = u_R(t)$ of class $C^{(2)}([a,b])$ which is (I_q)- or $(\hat{I}_q)$-stable in $[a,b]$. Then there exists an $\varepsilon_0 > 0$ such that for $0 < \varepsilon \leq \varepsilon_0$ the problem (DP_2) has a solution $y = y(t,\varepsilon)$ for t in $[a,b]$ which satisfies

$$|y(t,\varepsilon) - u_L(t)| \leq w_R(t,\varepsilon) + \Gamma(\varepsilon)$$

or

$$|y(t,\varepsilon) - u_R(t)| \leq w_L(t,\varepsilon) + \Gamma(\varepsilon).$$

Here, w_R and w_L are as given in Theorem 3.1 with u replaced by u_L and u_R, respectively, and $|\Gamma(\varepsilon)| \leq c\varepsilon^{1/(2q+1)}$ for the case of weak stability and for the case of locally weak stability with $q = 0$, while $|\Gamma(\varepsilon)| \leq c\varepsilon^{1/\{2q(2q+1)\}}$ for the case of locally weak stability with $q \geq 1$.

Proof: The proof will be similar to that of Theorem 3.1. We shall only consider the case of the reduced problem (R_L) having the solution u_L which is locally weakly stable and (I_q)-stable, since the other cases can be proved in a similar manner.

We define, for $a \leq t \leq b$ and $\varepsilon > 0$

$$\alpha(t,\varepsilon) = u_L(t) - w_R(t,\varepsilon) - \Gamma(\varepsilon)$$

$$\beta(t,\varepsilon) = u_L(t) + w_R(t,\varepsilon) + \Gamma(\varepsilon),$$

where $\Gamma(\varepsilon) = (\varepsilon^\rho\gamma/m)^{1/(2q+1)}$, $\rho = 1$ for weak stability and for locally weak stability with $q = 0$, $\rho = 1/(2q)$ for locally weak stability with $q \geq 1$, and $\gamma > 0$ will be made precise below. We recall that

$$w_R(t,\varepsilon) = \begin{cases} |B-u_L(b)|\exp[-(m/\varepsilon)^{1/2}(b-t)] & \text{if } q = 0, \\ |B-u_L(b)|\varepsilon^{1/2q}\,[\varepsilon^{1/2}+\sigma(b-t)]^{-1/q} & \text{if } q \geq 1, \end{cases}$$

where

$$\sigma = m^{1/2}\,q|B-u_L(b)|^q[(q+1)(2q+1)!]^{-1/2}.$$

It only remains to show that α and β satisfy the required differential inequalities. Consider then just the function β since the verification for α follows by symmetry. It follows from Taylor's Theorem and the hypothesis of (I_q)-stability that

$$\begin{aligned} F(t,\beta,\beta') - \varepsilon\beta'' &= \sum_{j=1}^{2q} \frac{1}{j!}\,\partial_y^j\, h(t,u_L)(\beta-u_L)^j \\ &\quad + \frac{1}{(2q+1)!}\,\partial_y^{2q+1}\, h(t,\eta)(\beta-u_L)^{2q+1} \\ &\quad + f(t,\beta)\,w_R' - \varepsilon u_L'' - \varepsilon w_R'' \\ &\geq \frac{\varepsilon^\rho\gamma}{(2q+1)!} - |f(t,\beta)w_R'| - \varepsilon|u_L''|. \end{aligned}$$

Thus, if u_L is weakly stable, i.e., if

$$f(t,\xi)w_R' \geq 0,$$

we have

$$\begin{aligned} F(t,\beta,\beta') - \varepsilon\beta'' &\geq \frac{\varepsilon^\rho\gamma}{(2q+1)!} - \varepsilon|u_L''| \\ &\geq 0 \end{aligned}$$

if we choose $\gamma \geq |u_L''|(2q+1)!$. On the other hand, if u_L is only locally weakly stable, then

$$f(t,\xi)w_R' \geq 0 \quad \text{in} \quad [b-\delta/2,b],$$

but in the rest of the interval $[a,b]$,

$$w_R' = \begin{cases} |B-u_L(b)|(m/\varepsilon)^{1/2}\exp[-(m/\varepsilon)^{1/2}(b-t)], & \text{if } q = 0, \\ |B-u_L(b)|\varepsilon^{1/2q}\,\sigma q^{-1}[\varepsilon^{1/2} + \sigma(b-t)]^{-(q+1)/q}, & \text{if } q \geq 1 \end{cases}$$

so that, for sufficiently small ε,

$$|f(t,\xi)w_R'| \leq \begin{cases} c_1\varepsilon, & \text{if } q = 0 \\ c_1\,\varepsilon^{1/2q}, & \text{if } q \geq 1 \end{cases}$$

for some $c_1 > 0$. Thus

$$F(t,\beta,\beta') - \varepsilon\beta'' \geq 0$$

if we choose $\gamma \geq (c_1 + |u_L''|)(2q+1)!$. The conclusion of the theorem follows from Theorem 2.1.

Theorem 4.3. Assume that the reduced problem (R_L) or (R_R) has a weakly or locally weakly stable solution $u = u_L(t)$ or $u = u_R(t)$ of class $C^{(2)}([a,b])$ which is also (II_n)- or $(\tilde{II}_n)$-stable in $[a,b]$. Assume also that $u_L'' \geq 0$ in (a,b) and $u_L(b) \leq B$ or $u_R'' \geq 0$ in (a,b) and $u_R(a) \leq A$. Then there exists an $\varepsilon_0 > 0$ such that for $0 < \varepsilon \leq \varepsilon_0$ the problem (DP_2) has a solution $y = y(t,\varepsilon)$ for t in $[a,b]$ which satisfies

$$0 \leq y(t,\varepsilon) - u_L(t) \leq w_R(t,\varepsilon) + \Gamma(\varepsilon)$$

or

$$0 \leq y(t,\varepsilon) - u_R(t) \leq w_L(t,\varepsilon) + \Gamma(\varepsilon).$$

Here w_R and w_L are as given in Theorem 3.2 with u replaced by u_L and u_R, respectively, and

$$|\Gamma(\varepsilon)| \leq \begin{cases} c\varepsilon^{1/n} & \text{for the weakly stable case,} \\ c\varepsilon^{1/\{n(n-1)\}} & \text{for the locally weakly stable case.} \end{cases}$$

Proof: Let us consider only the case of u_R. We can prove this case by defining

$$\alpha(t,\varepsilon) = u_R(t),$$

$$\beta(t,\varepsilon) = u_R(t) + w_L(t,\varepsilon) + (\varepsilon^{\rho}\gamma m^{-1})^{1/n},$$

where $\rho = 1$ (or $\rho = 1/(n-1)$), if u_R is weakly (or locally weakly) stable and proceeding as in the proof of Theorem 4.2.

Theorem 4.4. Assume that the reduced problem (R_L) or (R_R) has a weakly or locally weakly stable solution $u = u_L(t)$ or $u = u_R(t)$ of class $C^{(2)}([a,b])$ which is also (III_n)- or $(\tilde{III}_n)$-stable. Assume also that $u_L'' \leq 0$ in (a,b) and $u_L(b) \geq B$ or $u_R'' \leq 0$ in (a,b) and $u_R(a) \geq A$. Then there exists an $\varepsilon_0 > 0$ such that for $0 < \varepsilon \leq \varepsilon_0$ the problem (DP_2) has a solution $y = y(t,\varepsilon)$ for t in $[a,b]$ which satisfies

$$-w_R(t,\varepsilon) - \Gamma(\varepsilon) \leq y(t,\varepsilon) - u_L(t) \leq 0$$

or

$$-w_L(t,\varepsilon) - \Gamma(\varepsilon) \leq y(t,\varepsilon) - u_R(t) \leq 0,$$

where w_R and w_L are as given in Theorem 3.2 with u replaced by u_L and u_R, respectively, and Γ is as given in Theorem 4.3.

Proof: Simply let $y \to -y$ and apply Theorem 4.3.

We next consider the solution $u = u(t)$ of the reduced equation (R) which does not satisfy either boundary condition but which is locally strongly or locally weakly stable. The proofs of the following two theorems are only a slight modification of the proofs of Theorems 4.2 and 4.3.

Theorem 4.5. Let the reduced equation (R) have a locally strongly or locally weakly stable solution $u = u(t)$ of class $C^{(2)}([a,b])$ which is also (I_q)- or $(\tilde{I}_q)$-stable in $[a,b]$. Then there exists an $\varepsilon_0 > 0$ such that for $0 < \varepsilon \leq \varepsilon_0$ the problem (DP_2) has a solution $y = y(t,\varepsilon)$ for t in $[a,b]$ which satisfies

$$|y(t,\varepsilon) - u(t)| \leq w_L(t,\varepsilon) + w_R(t,\varepsilon) + \Gamma(\varepsilon).$$

Here, Γ is as given in Theorem 4.2,

$$w_L = |A-u(a)|\exp[\lambda(t-a)], \quad w_R = |B-u(b)|\exp[\lambda(b-t)],$$

and $\lambda = -k/\varepsilon + O(1)$, if $u(t)$ is locally strongly stable, while w_L, w_R are as given in Theorem 3.1, if $u(t)$ is locally weakly stable.

Theorem 4.6. Let the reduced equation (R) have a locally strongly or locally weakly stable solution $u = u(t)$ of class $C^{(2)}([a,b])$ which is also (II_n)- or $(\tilde{II}_n)$-stable in $[a,b]$. Assume also that $u'' \geq 0$ in (a,b), $u(a) \leq A$ and $u(b) \leq B$. Then there exists an $\varepsilon_0 > 0$ such that for $0 < \varepsilon \leq \varepsilon_0$ the problem (DP_2) has a solution $y = y(t,\varepsilon)$ for t in $[a,b]$ which satisfies

$$0 \le y(t,\varepsilon) - u(t) \le w_L(t,\varepsilon) + w_L(t,\varepsilon) + c\varepsilon^{1/n},$$

where w_L, w_R are as given in Theorem 4.5, if u is locally strongly stable, while w_L, w_R are as given in Theorem 3.2, if u is locally weakly stable.

If the function u is locally strongly or locally weakly stable and (III_n)- or $(\tilde{III}_n)$-stable, then the result analogous to Theorem 4.6 holds, provided that $u'' \le 0$ in (a,b), $u(a) \ge A$ and $u(b) \ge B$. We leave its precise formulation to the reader.

§4.2. Robin Problems: Boundary Layer Phenomena

We now turn to the occurrence of boundary layer phenomena for solutions of the Robin problems

$$\begin{aligned} &\varepsilon y'' = f(t,y)y' + g(t,y) = F(t,y,y'), \quad a < t < b, \\ &y(a,\varepsilon) - p_1 y'(a,\varepsilon) = A, \quad y(b,\varepsilon) = B, \end{aligned} \tag{RP_3}$$

$$\begin{aligned} &\varepsilon y'' = f(t,y)y' + g(t,y), \quad a < t < b, \\ &y(a,\varepsilon) - p_1 y'(a,\varepsilon) = A, \quad y(b,\varepsilon) + p_2 y'(b,\varepsilon) = B, \end{aligned} \tag{RP_4}$$

with their associated reduced problems

$$\begin{aligned} &f(t,u)u' + g(t,u) = 0, \quad a < t < t_1 \le b, \\ &u(a) - p_1 u'(a) = A, \end{aligned} \tag{$\tilde{R}_L$}$$

$$\begin{aligned} &f(t,u)u' + g(t,u) = 0, \quad a \le t_2 < t < b, \\ &u(b) = B, \end{aligned} \tag{R_R}$$

$$\begin{aligned} &f(t,u)u' + g(t,u) = 0, \quad a \le t_2 < t < b, \\ &u(b) + p_2 u'(b) = B, \end{aligned} \tag{$\tilde{R}_R$}$$

and

$$f(t,u)u' + g(t,u) = 0, \quad a < t < b. \tag{R}$$

(We note that the related problem $\varepsilon y'' = f(t,y)y' + g(t,y)$, $a < t < b$, with $y(a,\varepsilon)$, $y(b,\varepsilon) + p_2 y'(b,\varepsilon)$ prescribed, can be handled by making the change of variable $t \to a + b - t$ and applying the results for (RP_3) to the transformed problem.) We denote the solutions of $(\tilde{R}_L)$ and (R) by u_L and u, respectively, while the solutions of (R_R) and $(\tilde{R}_R)$ will be denoted, without any distinction, by u_R. Earlier definitions of stability apply to these u_L, u_R and u except that for the case (RP_3), the functions d_R, d and d_1 are uniformly small in $[a,b-\delta]$ (that is, $d_R, d, d_1 \equiv \delta$)) while for the case (RP_4) the functions d_L, d_R, d, d_1, and d_2 are uniformly small in $[a,b]$.

To provide some insight into the results for such problems, we consider three simple examples. The first two problems are

$$\varepsilon y'' = \pm y', \quad 0 < t < 1,$$
$$y(0,\varepsilon) - y'(0,\varepsilon) = 1, \quad y(1,\varepsilon) = 2. \qquad (E_{10}^{\pm})$$

The exact solution of (E_{10}^{-}) is, with $\Delta = e^{-1/\varepsilon} - 1 - \varepsilon^{-1}$,

$$y(t,\varepsilon) = \Delta^{-1}[e^{-1/\varepsilon} - 2(1+\varepsilon^{-1}) + e^{-t/\varepsilon}] \sim 2 - \varepsilon e^{-t/\varepsilon},$$

and therefore

$$\lim_{\varepsilon\to 0+} y(t,\varepsilon) = 2 \quad \text{in} \quad [0,1].$$

The limiting function $u_R \equiv 2$ is the solution of the reduced problem $u' = 0$, $u(1) = 2$. On the other hand, the exact solution of (E_{10}^{+}) is, with $\Delta = 1 - e^{-1/\varepsilon}(1-\varepsilon^{-1})$,

$$y(t,\varepsilon) = \Delta^{-1}[1 - 2e^{-1/\varepsilon}(1-\varepsilon^{-1}) + e^{-(1-t)/\varepsilon}]$$
$$\sim 1 + e^{-(1-t)/\varepsilon},$$

and so

$$\lim_{\varepsilon\to 0+} y(t,\varepsilon) = 1 \quad \text{for} \quad 0 \le t \le 1-\delta < 1.$$

Here the limiting function $u_L \equiv 1$ is of course the solution of the reduced problem $u' = 0$, $u(0) - u'(0) = 1$.

The third problem is

$$\varepsilon y'' = y', \quad 0 < t < 1,$$
$$y(0,\varepsilon) - y'(0,\varepsilon) = 1, \quad y(1,\varepsilon) + y'(1,\varepsilon) = 2. \qquad (E_{11})$$

The exact solution is, with $\Delta = \varepsilon^{-1} - e^{1/\varepsilon}(1-\varepsilon^{-1})$,

$$y(t,\varepsilon) = \Delta^{-1}[\varepsilon^{-1} - 2e^{-1/\varepsilon}(1-\varepsilon^{-1}) + e^{-(1-t)/\varepsilon}]$$
$$\sim 1 + \varepsilon e^{-(1-t)/\varepsilon},$$

and consequently

$$\lim_{\varepsilon\to 0+} y(t,\varepsilon) = 1 \quad \text{for} \quad 0 \le t \le 1.$$

Here the limiting function $u_L \equiv 1$ is the solution of $u' = 0$, $u(0) - u'(0) = 1$, and in contrast with (E_{10}^{+}), is the uniform limit of the solution $y(t,\varepsilon)$ in the whole interval $[0,1]$.

We return now to the problem (RP_3) to show the relation between its solution and the solutions of the reduced problems $(\tilde{R}_L)$ and (R).

Theorem 4.7. Assume that the reduced problem (R) has a strongly or locally strongly stable solution $u = u(t)$ of class $C^{(2)}([a,b])$ which is also (I_q)- or $(\tilde{I}_q)$-stable in $[a,b]$. Then there exists an $\varepsilon_0 > 0$ such that for $0 < \varepsilon \le \varepsilon_0$ the problem (RP_3) has a solution $y = y(t,\varepsilon)$ for t in $[a,b]$ which satisfies

$$|y(t,\varepsilon) - u_L(t)| \le v_L(t,\varepsilon) + w_R(t,\varepsilon) + c\varepsilon^{1/(2q+1)},$$

where

$$w_R(t,\varepsilon) = |B - u(b)|\exp[-k\varepsilon^{-1}(b-t)]$$

and

$$v_L(t,\varepsilon) = \varepsilon(p_1k)^{-1}|A-u(a)+p_1u'(a)|\exp[-k\varepsilon^{-1}(t-a)].$$

Proof: We consider only the case of (I_q)-stability, and we define for $a \le t \le b$ the functions

$$\alpha(t,\varepsilon) = u(t) - v_L(t,\varepsilon) - w_L(t,\varepsilon) - (\varepsilon\gamma/m)^{1/(2q+1)}$$

$$\beta(t,\varepsilon) = u(t) + v_L(t,\varepsilon) + w_R(t,\varepsilon) + (\varepsilon\gamma/m)^{1/(2q+1)},$$

where $\gamma > 0$ will be determined later. Clearly, we have $\alpha(t,\varepsilon) \le \beta(t,\varepsilon)$, $\alpha(b,\varepsilon) \le B \le \beta(b,\varepsilon)$ and $\alpha(a,\varepsilon) - p_1\alpha'(a,\varepsilon) \le A \le \beta(a,\varepsilon) - p_1\beta'(a,\varepsilon)$, if ε is sufficiently small. We can show that $\varepsilon\alpha'' \ge F(t,\alpha,\alpha')$ and $\varepsilon\beta'' \le F(t,\beta,\beta')$. For example, if we write $h(t,y) = f(t,y)u' + g(t,y)$, we obtain

$$\begin{aligned}\varepsilon\alpha'' - F(t,\alpha,\alpha') &= \varepsilon u'' - \varepsilon v_L'' - \varepsilon w_R'' - \sum_{j=1}^{2q} \frac{1}{j!}\,\partial_y^j h(t,u)(\alpha-u)^j \\ &\quad - \frac{1}{(2q+1)!}\,\partial_y^{2q+1}h(t,\eta)(\alpha-u)^{2q+1} - f(t,\xi)(\alpha'-u') \\ &\ge \varepsilon|u''| - \varepsilon v_L'' - \varepsilon w_R'' + \frac{m}{(2q+1)!}\{v_L^{2q+1}+w_R^{2q+1}\} + \frac{\varepsilon\gamma}{(2q+1)!} \\ &\quad + f(t,\xi)(v_L' + w_R'), \\ &\ge \varepsilon|u''| - \varepsilon v_L'' - \varepsilon w_R'' + \frac{\varepsilon\gamma}{(2q+1)!} + f(t,\xi)(v_L' + w_R').\end{aligned}$$

Now, for t in $[a, a+\delta/2]$

$$f(t,\xi) \le -k < 0,$$

and hence,

$$f(t,\xi)v_L' \ge -kv_L' = \varepsilon v_L''.$$

Also, for sufficiently small ε, the term $-\varepsilon w_R'' + fw_R' = \tau(\varepsilon)$ is such that $|\tau| \le c_0\varepsilon$, for some $c_0 > 0$ and so in $[a,a+\delta/2]$,

$$\begin{aligned}\varepsilon\alpha'' - F(t,\alpha,\alpha') &\ge -\varepsilon|u''| + \frac{\varepsilon\gamma}{(2q+1)!} - c_0\varepsilon,\\ &\ge 0,\end{aligned}$$

if we choose $\gamma = (c_0 + |u''|)(2q+1)!$. Similarly, for t in $[b-\delta/2,b]$, we have

$$f(t,\xi) \ge k > 0,$$

and so

$$f(t,\xi)w_R' \ge kw_R' = \varepsilon w_R''.$$

For sufficiently small ε, the term $-\varepsilon v_L'' + fv_L' = \tilde{\tau}(\varepsilon)$ satisfies $|\tilde{\tau}| \le c_1\varepsilon$ for some $c_1 > 0$. Thus, in $[b-\delta/2,b]$,

$$\begin{aligned}\varepsilon\alpha'' - F(t,\alpha,\alpha') &\ge -\varepsilon|u''| + \frac{\varepsilon\gamma}{(2q+1)!} - c_1\varepsilon\\ &\ge 0,\end{aligned}$$

if we choose $\gamma \ge (c_1 + |u''|)(2q+1)!$. Finally, in $[a+\delta/2, b-\delta/2]$, for sufficiently small ε, we have

$$\begin{aligned}\varepsilon\alpha'' - F(t,\alpha,\alpha') &\ge -\varepsilon|u''| + \frac{\varepsilon\gamma}{(2q+1)!} + \tau(\varepsilon) + \tilde{\tau}(\varepsilon)\\ &\ge 0,\end{aligned}$$

by choosing $\gamma \ge \{c_1 + c_0 + |u''|\}(2q+1)!$.

The result of the theorem now follows from Theorem 2.3.

Following the proofs given above and in Theorem 4.3, we can obtain the next result.

Theorem 4.8. Assume that the reduced problem (R) has a strongly or locally strongly stable solution $u = u(t)$ of class $C^{(2)}([a,b])$ which is also (II_n)- or $(\tilde{II}_n)$-stable in $[a,b]$. Assume also that $u'' \ge 0$ in (a,b), $u(a) - p_1u'(a) \le A$ and $u(b) \le B$. Then there exists an $\varepsilon_0 > 0$ such that for $0 < \varepsilon \le \varepsilon_0$ the problem (RP_3) has a solution $y = y(t,\varepsilon)$ for t in $[a,b]$ which satisfies

$$0 \leq y(t,\varepsilon) - u(t) \leq v_L(t,\varepsilon) + w_R(t,\varepsilon) + c\varepsilon^{1/n},$$

where w_R and v_L are as given in Theorem 4.7.

On the other hand, if the strongly or locally strongly stable solution $u(t)$ of the reduced problem (R) is (III_n)- or $(I\tilde{I}I_n)$-stable and satisfies $u'' \leq 0$ in (a,b), $u(a) - p_1u'(a) \geq A$ and $u(b) \geq B$, then there exists an $\varepsilon_0 > 0$ such that for $0 < \varepsilon \leq \varepsilon_0$ the problem (RP_3) has a solution $y(t,\varepsilon)$ in $[a,b]$ satisfying

$$-v_L(t,\varepsilon) - w_R(t,\varepsilon) - c\varepsilon^{1/n} \leq y(t,\varepsilon) - u(t) \leq 0.$$

The following results for the weakly or locally weakly stable solution $u(t)$ can be proved using analogous arguments.

Theorem 4.9. Let the reduced problem (R) have a weakly or locally weakly stable solution $u = u(t)$ of class $C^{(2)}([a,b])$ which is also (I_q)- or $(\tilde{I}_q)$-stable in $[a,b]$. Then there exists an $\varepsilon_0 > 0$ such that for $0 < \varepsilon \leq \varepsilon_0$ the problem (RP_3) has a solution $y = y(t,\varepsilon)$ for t in $[a,b]$ which satisfies

$$|y(t,\varepsilon) - u(t)| \leq v_L(t,\varepsilon) + w_R(t,\varepsilon) + \Gamma(\varepsilon),$$

where w_R is as given in Theorem 3.1 and v_L in Theorem 3.4.

Let the reduced problem (R) have a weakly or locally weakly stable solution $u = u(t)$ of class $C^{(2)}([a,b])$ which is also (II_n)- or $(I\tilde{I}_n)$-stable in $[a,b]$, and, moreover, satisfies $u'' \geq 0$ in (a,b), $u(a) - p_1u'(a) \leq A$ and $u(b) \leq B$. Then there exists an $\varepsilon_0 > 0$ such that for $0 < \varepsilon \leq \varepsilon_0$ the problem (RP_3) has a solution $y = y(t,\varepsilon)$ for t in $[a,b]$ which satisfies

$$0 \leq y(t,\varepsilon) - u(t) \leq v_L(t,\varepsilon) + w_R(t,\varepsilon) + \tilde{\Gamma}(\varepsilon),$$

where w_R is as given in Theorem 3.2 and v_L in Theorem 3.5.

We leave it to the reader to formulate the statement of the results for a weakly or locally weakly stable solution $u(t)$ which is also (III_n)- or $(\tilde{III}_n)$-stable.

The following results follow from the proofs of Theorems 4.7 - 4.9.

Corollary 4.2. Suppose the reduced problem $(\tilde{R}_L)$ has a solution $u = u_L(t)$ satisfying the hypotheses in each of the Theorems 4.7 - 4.9, then the conclusions of each of the theorems hold with u_L replacing u, but without the term $v_L(t,\varepsilon)$.

On the other hand, if the reduced problem (R_R) has a solution $u = u_R(t)$ satisfying the hypotheses in each of the Theorems 4.7 - 4.9, then the conclusions of each of the theorems hold with u_R replacing u, but without the term $w_R(t,\varepsilon)$.

With these results, it is now an easy matter to deal with the problem (RP_4), and so we omit the proofs of the following results.

Theorem 4.10. Assume that the reduced equation (R) has a locally strongly (or weakly) stable solution $u = u(t)$ of class $C^{(2)}([a,b])$ which is also (I_q)-stable in $[a,b]$. Then there exists an $\varepsilon_0 > 0$ such that for $0 < \varepsilon \leq \varepsilon_0$ the problem (RP_4) has a solution $y = y(t,\varepsilon)$ for t in $[a,b]$ which satisfies

$$|y(t,\varepsilon) - u(t)| \leq v_L(t,\varepsilon) + v_R(t,\varepsilon) + \Gamma(\varepsilon).$$

Here

$$v_L = \varepsilon(p_1k)^{-1}|A-u(a)+p_1u'(a)|\exp[-k\varepsilon^{-1}(t-a)],$$

$$v_R = \varepsilon(p_2k)^{-1}|B-u(b)-p_2u'(b)|\exp[-k\varepsilon^{-1}(b-t)],$$

if u is locally strongly stable, and v_L, v_R are as given in Theorem 3.7 if u is locally weakly stable, and Γ is as given in Theorem 4.2.

Theorem 4.11. Assume that the reduced equation (R) has a locally strongly (or weakly) stable solution $u = u(t)$ of class $C^{(2)}([a,b])$ which is also (II_n)-stable in $[a,b]$. Assume also that $u'' \geq 0$ in $[a,b]$, $u(a) - p_1u'(a) \leq A$ and $u(b) + p_2u'(b) \leq B$. Then there exists an $\varepsilon_0 > 0$ such that for $0 < \varepsilon \leq \varepsilon_0$ the problem (RP_4) has a solution $y = y(t,\varepsilon)$ for t in $[a,b]$ which satisfies

$$0 \leq y(t,\varepsilon) - u(t) + v_L(t,\varepsilon) + v_R(t,\varepsilon) + \Gamma(\varepsilon),$$

where v_L and v_R are as given in Theorem 4.10 (or Theorem 3.8) if u is locally strongly (or weakly) stable, and Γ is as given in Theorem 4.3.

Corollary 4.3. Suppose the reduced problem $(\tilde{R}_L)$ has a solution $u = u_L(t)$ satisfying the hypotheses in each of the Theorems 4.9, 4.10. Then the conclusions of each of the theorems hold with u_L replacing u, but without the term $v_L(t,\varepsilon)$.

On the other hand, if the reduced problem $(\tilde{R}_R)$ has a solution $u = u_R(t)$ satisfying the hypotheses in each of the Theorems 4.9, 4.10, then the conclusions of each of the theorems hold with u_R replacing u, but without the term $v_R(t,\varepsilon)$.

The corresponding result for (III_n)-stable cases we leave to the reader to state.

§4.3. Interior Layer Phenomena

The remaining part of this chapter will be devoted to a discussion of interior crossing phenomena for solutions of the problems (DP_2), (RP_3) and (RP_4). What we have in mind is the following. Suppose, for example, that the reduced problem (R_L) has a solution $u = u_L(t)$ and the reduced problem (R_R) has a solution $u = u_R(t)$ which intersect at a point t_0 in (a,b) with unequal slopes, that is, $u_L(t_0) = u_R(t_0)$ and $u_L'(t_0) \neq u_R'(t_0)$. Then we have the reduced path

$$u_0(t) = \begin{cases} u_L(t), & a \leq t \leq t_0, \\ u_R(t), & t_0 \leq t \leq b, \end{cases} \tag{R_0}$$

where $u_0'(t_0^-) \neq u_0'(t_0^+)$. We ask under what conditions will the problem (DP_2) possess a solution $y = y(t,\varepsilon)$ such that

$$\lim_{\varepsilon\to 0+} y(t,\varepsilon) = u_0(t)? \tag{4.4}$$

The answer to this question is not as straightforward as it was in the case of the problem (DP_1) considered in the previous chapter. First of all, if the pair of reduced solutions u_L and u_R are both strongly stable, then such a pair can *never* attract a sol ution $y(t,\varepsilon)$ of (DP_2) in the sense of (4.4). The reason for this is simply that if the pair u_L, u_R crosses at t_0, then in the small interval $t_0-\delta \leq t \leq t_0+\delta$, the Cauchy problem

$$f(t,u)u' + g(t,u) = 0,$$

$$u(t_0) = \sigma(= u_L(t_0) = u_R(t_0))$$

must have two distinct solutions, but this is impossible in view of the smoothness of f and g and the fact that $|f| > 0$. Consequently, we can only expect the relation (4.4) from the crossing of two weakly stable solutions of the reduced equation (R) since if $u_L(t_0) = u_R(t_0) = \sigma$, then it necessarily follows that $f(t_0,\sigma) = 0$. With these remarks, we can now discuss some general results for this crossing phenomenon, first for the Dirichlet problem (DP_2) and then for the Robin problems (RP_3) and (RP_4).

To give our first results, let us recall from Theorem 3.9 the function

$$v_I(t,\varepsilon) = \begin{cases} \frac{1}{2}(\varepsilon/m)^{1/2}|u'(t_0^+)-u'(t_0^-)|\exp[-(\varepsilon/m)^{-1/2}|t-t_0|], & \text{if } q = 0, \\ \frac{1}{2}\,\sigma[1+q(\varepsilon(2q+2)!/2m)^{-1/2}\sigma^q|t-t_0|]^{-1/q} & , \text{ if } q \geq 1, \end{cases}$$

where

$$\sigma^{q+1} = |u'(t_0^+) - u'(t_0^-)|\{\varepsilon(2q+2)!/2m\}^{1/2}.$$

Theorem 4.12. Assume that the reduced path (R_0) is of class C^2, weakly stable and (I_q)-stable in $[a,b]$. Then there exists an $\varepsilon_0 > 0$ such that for $0 < \varepsilon \leq \varepsilon_0$ the problem (DP_2) has a solution $y = y(t,\varepsilon)$ for t in $[a,b]$ which satisfies

$$|y(t,\varepsilon) - u_0(t)| \leq v_1(t,\varepsilon) + c\varepsilon^{1/(2q+1)},$$

where $v_1(t,\varepsilon)$ is $v_I(t,\varepsilon)$ with $|u'(t_0^+) - u'(t_0^-)|$ replaced by $|u_R'(t_0) - u_L'(t_0)|$.

Proof: The proof is similar to that of Theorem 3.9. We can suppose that $u_L'(t_0) < u_R'(t_0)$, and define for $a \leq t \leq b$ and $\varepsilon > 0$

$$\alpha(t,\varepsilon) = u_0(t) - (\varepsilon\gamma m^{-1})^{1/(2q+1)},$$

$$\beta(t,\varepsilon) = u_0(t) + v_1(t,\varepsilon) + (\varepsilon\gamma m^{-1})^{1/(2q+1)},$$

where $\gamma \geq |u_0''|(2q+1)!$. We note that α is not differentiable at $t = t_0$; indeed, $\alpha'(t_0^-) < \alpha'(t_0^+)$. Nevertheless, by our choice of γ, we can show as in the proof of Theorem 3.9 that

$$\varepsilon\alpha'' \geq F(t,\alpha,\alpha') \quad \text{in} \quad (a,b) \setminus \{t_0\},$$

that is, α is a lower solution. As regards β, we see that by our choice of v_1, β is differentiable at $t = t_0$, and again by our choice of γ,

$$\varepsilon\beta'' \leq F(t,\beta,\beta') \quad \text{in} \quad (a,b) \setminus \{t_0\}.$$

Thus the results follow from Theorem 2.2.

Theorem 4.13. Assume that the reduced path (R_0) is of class C^2, weakly stable and (II_n)-stable. Moreover, assume $u_L'' \geq 0$, $u_R'' \geq 0$ and $u_L'(t_0) < u_R'(t_0)$. Then there exists an $\varepsilon_0 > 0$ such that for $0 < \varepsilon \leq \varepsilon_0$ the problem (DP_2) has a solution $y = y(t,\varepsilon)$ for t in $[a,b]$ which satisfies

$$0 \le y(t,\varepsilon) - u_0(t) \le v_1(t,\varepsilon) + c\varepsilon^{1/n},$$

where $v_1(t,\varepsilon)$ is $v_I(t,\varepsilon)$ with $\frac{1}{2}(n-1)$ in place of q and with $|u_R'(t_0) - u_L'(t_0)|$ in place of $|u'(t_0^+) - u'(t_0^-)|$.

Proof: This result follows by arguing as in the proof of the previous theorem if we define for $a \le t \le b$ and $\varepsilon > 0$

$$\alpha(t,\varepsilon) = u_0(t),$$

$$\beta(t,\varepsilon) = u_0(t) + v_1(t,\varepsilon) + (\varepsilon\gamma m^{-1})^{1/n},$$

where $\gamma \ge |u_0''|n!$ and

$$v_1(t,\varepsilon) = \frac{1}{2}\sigma[1 + \frac{1}{2}(n-1)(\varepsilon(n+1)!/2m)^{-1/2}\sigma^{1/2(n-1)}|t-t_0|]^{-2/(n-1)},$$

with

$$\sigma^{n+1} = \varepsilon(n+1)!|u_R'(t_0) - u_L'(t_0)|/2m.$$

If the weakly stable reduced path R_0 is (III_n)-stable, then a result analogous to Theorem 4.13 is valid provided that $u_0'' \le 0$ in $(a,t_0) \cup (t_0,b)$ and $u_L'(t_0) > u_R'(t_0)$. We leave its precise formulation to the reader.

The above results can be generalized to the three-branch reduced path

$$\tilde{u}_0(t) = \begin{cases} u_L(t), & a \le t \le t_1, \\ u(t), & t_1 \le t \le t_2, \\ u_R(t), & t_2 \le t \le b, \end{cases}$$

where the solutions u_L, u_R intersect the middle solution u of the reduced equation (R), such that $u_L(t_1) = u(t_1)$, $u_L'(t_1) \ne u'(t_1)$, $u(t_2) = u_R(t_2)$ and $u'(t_2) \ne u_R'(t_2)$. If $t_1 = t_2$, this becomes the reduced path $u_0(t)$ above.

Theorem 4.14. Assume that the reduced path $\tilde{u}_0(t)$ is of class C^2, weakly stable and (I_q)-stable in $[a,b]$. Then there exists an $\varepsilon_0 > 0$ such that for $0 < \varepsilon \le \varepsilon_0$ the problem (DP_2) has a solution $y = y(t,\varepsilon)$ for t in $[a,b]$ which satisfies

$$|y(t,\varepsilon) - \tilde{u}_0(t)| \le v_1(t,\varepsilon) + v_2(t,\varepsilon) + c\varepsilon^{1/(2q+1)}.$$

Here $v_1(t,\varepsilon)$ is $v_I(t,\varepsilon)$ with t_0 and $|u'(t_0^+) - u'(t_0^-)|$ replaced, respectively, by t_1 and $|u'(t_1) - u_L'(t_1)|$, while $v_2(t,\varepsilon)$ is $v_I(t,\varepsilon)$ with t_0 and $|u'(t_0^+) - u'(t_0^-)|$ replaced, respectively, by t_2 and $|u_R'(t_2) - u'(t_2)|$.

The corresponding result for the (II_n)-stable solution $\tilde{u}_0(t)$ we will leave to the reader to formulate, but we wish to state the result for the following reduced path

$$u_1(t) = \begin{cases} u(t)\ , & a \le t \le t_2, \\ u_R(t), & t_2 \le t \le b, \end{cases}$$

where $u'(t_2) \ne u_R'(t_2)$. Here, since the reduced solution $u(t)$ does not satisfy the boundary condition at $t = a$, we expect that the solution of (DP_2) which follows the reduced path $u_1(t)$ will behave nonuniformly at $t = a$. Similar results hold for the "reflected path"

$$u_2(t) = \begin{cases} u_L(t), & a \le t \le t_1, \\ u(t)\ , & t_1 \le t \le b. \end{cases}$$

Theorem 4.15. Assume that the function $u(t)$ is locally strongly (or weakly) stable and that the reduced path $u_1(t)$ is of class C^2, weakly stable and (I_q)- or $(\tilde{I}_q)$-stable in $[a,b]$. Then there exists an $\varepsilon_0 > 0$ such that for $0 < \varepsilon \le \varepsilon_0$ the problem (DP_2) has a solution $y = y(t,\varepsilon)$ for t in $[a,b]$ which satisfies

$$|y(t,\varepsilon) - u_1(t)| \le w_L(t,\varepsilon) + v_2(t,\varepsilon) + c\varepsilon^{1/(2q+1)}.$$

Here, w_L is as given in Theorem 4.1 or Theorem 4.3, according as u is locally strongly or weakly stable, and v_2 is as given in Theorem 4.14.

We conclude this chapter with the following theorems for the Robin problems (RP_3) and (RP_4) whose proofs are similar to those of Theorems 4.12 - 4.15.

Theorem 4.16. Let the assumptions of Theorems 4.12 - 4.14 hold with the solution of (R_L) replaced by the solution of $(\tilde{R}_L)$ for the Robin problem (RP_3) or with the solutions of (R_L) and (R_R) replaced, respectively, by the solutions of $(\tilde{R}_L)$ and $(\tilde{R}_R)$ for the Robin problem (RP_4). Then the conclusions of Theorems 4.12 - 4.14 hold with (DP_2) replaced by (RP_3) or (RP_4).

Theorem 4.17. Let the assumptions of Theorem 4.15 hold for the Robin problem (RP_3) or with the solution of (R_R) replaced by the solution of $(\tilde{R}_R)$ for the Robin problem (RP_4). Then the conclusions of Theorem 4.15 hold with (DP_2) replaced by (RP_3) or (RP_4) and with w_L replaced by v_L. Here v_L is given in Theorem 4.7 (or Theorem 3.5) if u is locally strongly (or locally weakly) stable.

Notes and Remarks

4.1. The theory developed above applies with little change to the more general equation $\varepsilon y'' = f(t,y,\varepsilon)y' + g(t,y,\varepsilon)$ and boundary data $A(\varepsilon)$ and $B(\varepsilon)$, provided that

$$\{f,g\}\ (t,y,\varepsilon) = \{f,g\}\ (t,y,0) + o(1),$$

for (t,y) in the appropriate domain, and

$$\{A(\varepsilon),B(\varepsilon)\} = \{A(0),B(0)\} + o(1) \quad \text{for} \quad 0 < \varepsilon \leq \varepsilon_0.$$

4.2. The Dirichlet problem (DP_2) has been studied by many people including Tschen [85], von Mises [64], Oleinik and Zizina [70], Coddington and Levinson [14], Bris [7], Wasow [92], Vasil'eva [87], Erdélyi [22], Willett [96], O'Malley [72], Cole [55; Chapter 2], Eckhaus [21; Chapter 5], Cook and Eckhaus [16], Chang [9], [10], Dorr, Parter, Shampine [20], Ackerberg and O'Malley [1], Habets [28], Habets and Laloy [31], Kreiss and Parter [56], Matkowsky [63], Howes [39] and Olver [71]. The majority of these papers deal with solutions of the various reduced problems which are strongly stable and/or (I_0)-stable.

The corresponding Robin problems (RP_3) and (RP_4) have also been studied extensively. We mention only the papers by Bris [7], Vasil'eva [87], O'Malley [75; Chapter 7], Cohen [15], Keller [51], Macki [62], Searl [84], Dorr, Parter, Shampine [20], Habets and Laloy [31] and Howes [43].

4.3. The stability conditions given in Definitions 4.1 and 4.2 can be weakened in the following way, as first observed by Coddington and Levinson [14]. Consider first a solution $u = u_R(t)$ of the reduced problem (R_R). If there is a positive constant k such that $f(t,u_R(t)) \leq -k < 0$ in $[a,b]$ <u>and</u>

$$(u_R(a)-A) \int_{u_R(a)}^{\xi} f(a,s)ds > 0 \quad \text{for all} \quad \xi \quad \text{in} \quad (u_R(a),A] \quad \text{or} \quad [A,u_R(a)),$$

then the problem (DP_2) has a solution $y = y(t,\varepsilon)$ such that

$$\lim_{\varepsilon\to 0+} y(t,\varepsilon) = u_R(t) \quad \text{for} \quad a < a+\delta \leq t \leq b,$$

Similarly, if there is a positive constant k such that $f(t,u_L(t)) \geq k > 0$ in $[a,b]$, where $u = u_L(t)$ is a solution of the reduced problem (R_L), then the problem (DP_2) has a solution $y = y(t,\varepsilon)$ such that

$$\lim_{\varepsilon\to 0+} y(t,\varepsilon) = u_L(t) \quad \text{for} \quad a \leq t \leq b-\delta < b \quad \text{provided that}$$

$$(u_L(b)-B)\int_{u_L(b)}^{\eta} f(b,s)ds < 0 \quad \text{for all} \quad \eta \quad \text{in} \quad (u_L(b),B]$$

$$\text{or} \quad [B,u_L(b)).$$

4.4. In considering the asymptotic behavior of solutions of (DP_2), (RP_3) and (RP_4) in the presence of locally strongly or weakly or locally weakly stable solutions of the corresponding reduced problems, we always assumed an additional form of y-stability. If such y-stability is not assumed then the theory becomes incredibly complicated, even in the linear case, that is, $f(t,y) = f(t)$ and $g(t,y) = g_0(t)y$. The reader can consult the book of O'Malley [75; Chapter 8] and the article of Olver [71] for discussions of such problems in the linear case. A corresponding nonlinear theory is nonexistent, in the sense that if our assumptions of y-stability are dropped, then no conclusions regarding the existence and the behavior of solutions of these problems can be drawn.

4.5. It is also possible to weaken our assumptions regarding the Robin problems (RP_3) and (RP_4) in the case that the reduced solution is strongly stable, as was first observed by Bris [7]. Suppose, for example, that we consider the problem (RP_4) and let $u = u_L(t)$ be a strongly stable solution of the reduced problem (R_L). Then if $f(a,u_L(a)) + p_1h_y(a,u_L(a)) \neq 0$, for $h(t,y) = f(t,y)u_L'(t) + g(t,y)$, the problem (RP_4) has a solution $y = y(t,\varepsilon)$ such that

$$\lim_{\varepsilon\to 0+} y(t,\varepsilon) = u_L(t) \quad \text{for} \quad a \leq t \leq b.$$

Similarly, if $u = u_R(t)$ is a strongly stable solution of the reduced problem $(\tilde{R}_R)$ such that $f(b,u_R(b)) - p_2h_y(b,u_R(b)) \neq 0$, then the problem (RP_4) has a solution $y = y(t,\varepsilon)$ such that

$$\lim_{\varepsilon\to 0+} y(t,\varepsilon) = u_R(t) \quad \text{for} \quad a \leq t \leq b\,.$$

Note that if u_L or u_R is (I_q)-, (II_n)- or (III_n)-stable then these two conditions are automatically satisfied, since p_1 and p_2 are positive and $h_y \geq 0$.

4.6. We have not considered shock layer behavior for solutions of (DP_2), (RP_3) or (RP_4) in this chapter. The reader can consult Howes [39], [43] for a discussion of these phenomena.

Chapter V
Quadratic Singular Perturbation Problems

§5.1. Introduction

In this chapter we investigate the asymptotic behavior of solutions of boundary value problems for the differential equation

$$\varepsilon y'' = p(t,y)\, y'^2 + g(t,y), \quad a < t < b, \tag{DE}$$

The novelty here is the presence of the quadratic term in y'. The more general differential equation

$$\varepsilon y'' = p(t,y)y'^2 + f(t,y)y' + g(t,y)$$

will not be studied, since it can be reduced to the form (DE) in some cases by the familiar device of completing the square. Our decision to study the simpler equation (DE) rather than the more general equation stems from a desire to present representative results for this "quadratic" class of problems without having to deal with extra complexities in notation.

§5.2. The Dirichlet Problem: Boundary Layer Phenomena

We shall first consider the following Dirichlet problem

$$\begin{aligned} &\varepsilon y'' = p(t,y)y'^2 + g(t,y), \quad a < t < b, \\ &y(a,\varepsilon) = A, \quad y(b,\varepsilon) = B. \end{aligned} \tag{DP_3}$$

To motivate some of the results to follow, it is useful to pay attention to the results that have been obtained for the *model* problem (cf. [27])

$$\begin{aligned} &\varepsilon y'' = 1 - y'^2, \quad 0 < t < 1 \\ &y(0,\varepsilon) = A, \quad y(1,\varepsilon) = B. \end{aligned} \tag{E_{12}}$$

The exact solution of (E_{12}), for $|A-B| < 1$, as can be found by quadratures (cf. [94]), is

$$y(t,\varepsilon) = \varepsilon \ln \cosh[\{t - \tfrac{1}{2}(A-B+1)\}/\varepsilon] + \tfrac{1}{2}(A+B-1) + O(\varepsilon).$$

Since $\cosh z = \frac{1}{2}(e^z + e^{-z}) \sim \frac{1}{2} e^z$, as $z \to \infty$, it is obvious that the asymptotic behavior of the solution depends critically on the relative position of the boundary values A and B. In particular, we can distinguish the following three cases:

(i) If $|A-B| < 1$, then

$$\begin{aligned} \lim_{\varepsilon\to 0^+} y(t,\varepsilon) &= \lim_{\varepsilon\to 0^+} \varepsilon\ln[\tfrac{1}{2}\exp(\{t-\tfrac{1}{2}(A-B+1)\}/\varepsilon] + \tfrac{1}{2}(A-B+1) \\ &= |t - \tfrac{1}{2}(A-B+1)| + \tfrac{1}{2}(A-B+1) \\ &= \begin{cases} A-t, & 0 \le t \le t_0, \\ t+B-1, & t_0 \le t \le 1, \end{cases} \end{aligned}$$

where $t_0 = \frac{1}{2}(A-B+1)$.

(ii) If $A-B = -1$, then

$y(t,\varepsilon) = t+A = t+B-1$ is the solution of (E_{12}).

(iii) If $A-B = 1$, then

$y(t,\varepsilon) = A-t$ is the solution of (E_{12}).

On the other hand, if $|A-B| > 1$ then the exact solution of (E_{12}) can be found again by quadratures. There are two cases:

(iv) If $B-A > 1$, then

$$\lim_{\varepsilon\to 0^+} y(t,\varepsilon) = t+B-1 \quad \text{for } \delta \le t \le 1, \text{ where } 0 < \delta < 1.$$

(v) If $B-A < -1$, then

$$\lim_{\varepsilon\to 0^+} y(t,\varepsilon) = A-t \quad \text{for } 0 \le t \le 1-\delta.$$

In case (i) above, the solution $y(t,\varepsilon)$ follows the angular path $u_0(t) = \max\{A-t, t+B-1\}$ in $[0,1]$, while in cases (iv) and (v), the solution displays the familiar boundary layer behavior at $t = 0$ and

$t = 1$, respectively. The functions $u_L(t) = A-t$ and $u_R(t) = t+B-1$ are of course solutions of the reduced equation

$$1 - u'^2 = 0 \tag{R_{12}}$$

which satisfy, respectively, $u_L(0) = A$ and $u_R(1) = B$. However, the equation (R_{12}) has another pair of solutions with these properties, namely $\tilde{u}_L(t) = t+A$ and $\tilde{u}_R(t) = B+1-t$. As we have just seen, the functions $\tilde{u}_L$, $\tilde{u}_R$ do not participate in the asymptotic description of solutions of (E_{12}). Therefore, in any asymptotic theory for the general problem (DP_3), we should give criteria to distinguish possible limiting solutions from all solutions of the corresponding reduced problems.

As a first step toward developing such a theory, we define the following reduced problems

$$\begin{aligned} &p(t,u)u'^2 + g(t,u) = 0, \quad a < t < t_1 \le b,\\ &u(a) = A, \end{aligned} \tag{R_L}$$

$$\begin{aligned} &p(t,u)u'^2 + g(t,u) = 0, \quad a \le t_2 < t < b,\\ &u(b) = B, \end{aligned} \tag{R_R}$$

and

$$p(t,u)u'^2 + g(t,u) = 0, \quad a < t < b. \tag{R}$$

Solutions of (R_L), (R_R) and (R) will be denoted throughout this chapter by u_L, u_R, and u, respectively. In view of the quadratic nonlinearity in u', the reduced equation (R) may have, in addition to the general solution, a singular solution u_s (cf. [48; Chapter 3], [38]). This phenomenon of a singular solution did not arise in earlier chapters. A singular solution is easily visualized as the envelope of a one-parameter family of solutions of (R) and as such $u_s' \equiv 0$, that is, $u_s \equiv$ const. A simple example of such a situation is afforded by the Clairaut equation

$$u'^2 - u = 0, \tag{E_{13}}$$

whose singular solution $u_s = 0$ is the envelope of the family of solutions $u(t) = \frac{1}{4}(t+c)^2$; see Figure 5.1. Note that for the equation (E_{13}), if $u(t_0) = 0$, then $u(t) = \frac{1}{4}(t-t_0)^2$ and $u'(t_0) = 0$, that is, any member of the family $\{u(t)\}$ intersects the singular solution $u_s \equiv 0$ smoothly. Similarly, if a solution u of the equation (R) intersects a singular solution u_s at t_0, then $u(t_0) = u_s$ <u>and</u> $u'(t_0) = u_s'(t_0) = 0$, provided $p(t_0,u_s) \neq 0$ (cf. [48; Chapter 3]).

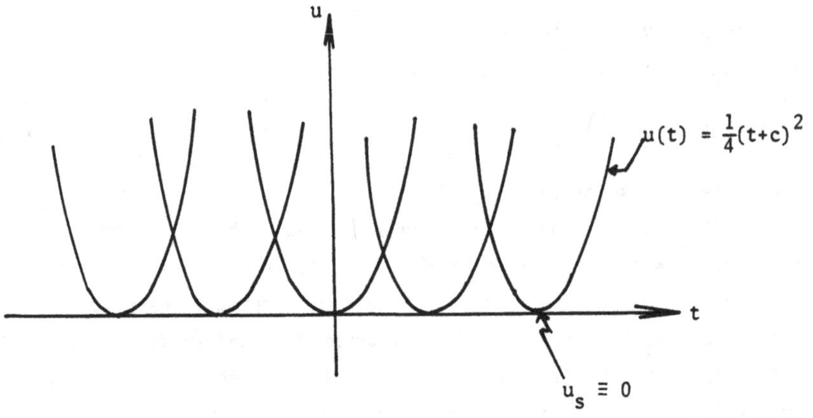

Figure 5.1

We now define the domains in which the functions p and g have certain properties relative to solutions of the reduced problems (R_L), (R_R) and (R).

Let $u = u_L(t)$ be a solution of (R_L) for $a \leq t \leq b$. If $u_L(b) \leq B$, we define the domain

$$\mathcal{R}^+(u_L) = \{(t,y) : a \leq t \leq b, \quad -\delta \leq y - u_L(t) \leq d_L^+(t)\},$$

where $d_L^+(t)$ is a positive continuous function satisfying

$$d_L^+(t) \equiv B - u_L(b) + \delta \quad \text{for} \quad b - \delta/2 \leq t \leq b$$

and

$$d_L^+(t) \equiv \delta \quad \text{for} \quad a \leq t \leq b - \delta.$$

If $u_L(b) \geq B$, we define the domain

$$\mathcal{R}^-(u_L) = \{(t,y) : a \leq t \leq b, \quad -d_L^-(t) \leq y - u_L(t) \leq \delta\},$$

where $d_L^-(t)$ is a positive continuous function satisfying

$$d_L^-(t) \equiv u_L(b) - B + \delta \quad \text{for} \quad b - \delta/2 \leq t \leq b$$

and

$$d_L^-(t) \equiv \delta \quad \text{for} \quad a \leq t \leq b - \delta.$$

Similarly, let $u = u_R(t)$ be a solution of (R_R) for $a \leq t \leq b$. If $u_R(a) \leq A$, we define the domain

$$R^+(u_R) = \{(t,y) : a \le t \le b, \ \ -\delta \le y - u_R(t) \le d_R^+(t)\},$$

where $d_R^+(t)$ is a positive continuous function satisfying

$$d_R^+(t) \equiv A - u_R(a) + \delta \quad \text{for} \quad a \le t \le a + \delta/2$$

and

$$d_R^+(t) \equiv \delta \quad \text{for} \quad a + \delta \le t \le b.$$

If $u_R(a) \ge A$, we define the domain

$$R^-(u_R) = \{(t,y) : a \le t \le b, \ \ -d_R^-(t) \le y - u_R(t) \le \delta\},$$

where $d_R^-(t)$ is a positive continuous function satisfying

$$d_R^-(t) \equiv u_R(a) - A + \delta \quad \text{for} \quad a \le t \le t + \delta/2$$

and

$$d_R^-(t) \equiv \delta \quad \text{for} \quad a + \delta \le t \le b.$$

Finally, for a solution $u = u(t)$ of (R), we define the domains $R^+(u)$, $R^-(u)$, $R^{\pm}(u)$, $R^{\mp}(u)$, depending on the relative values of $u(a)$, A, $u(b)$ and B.

If $u(a) \le A$ and $u(b) \le B$, we define the domain

$$R^+(u) = \{(t,y) : a \le t \le b, \ \ -\delta \le y - u(t) \le d^+(t)\},$$

where $d^+(t)$ is a positive continuous function satisfying

$$d^+(t) \equiv A-u(a)+\delta \quad \text{for} \quad a \le t \le a + \delta/2,$$

$$d^+(t) \equiv B-u(b)+\delta \quad \text{for} \quad b- \delta/2 \le t \le b,$$

and

$$d^+(t) \equiv \delta \qquad \text{for} \quad a+\delta \le t \le b-\delta.$$

If $u(a) \ge A$ and $u(b) \ge B$, we define the domain

$$R^-(u) = \{(t,y) : a \le t \le b, \ \ -d^-(t) \le y-u(t) \le \delta\},$$

where $d^-(t)$ is a positive continuous function satisfying

$$d^-(t) \equiv u(a) - A+\delta \quad \text{for} \quad a \le t \le a + \delta/2,$$

$$d^-(t) \equiv u(b) - B+\delta \quad \text{for} \quad b - \delta/2 \le t \le b,$$

and

$d^-(t) \equiv \delta$ for $a+\delta \leq t \leq b-\delta$.

If $u(a) \leq A$ and $u(b) \geq B$, we define the domain

$$R^{\pm}(u) = \{(t,y) : a \leq t \leq b, \ -d^-(t) \leq y-u(t) \leq d^+(t)\},$$

where d^+, d^- are positive continuous functions satisfying

$$d^+(t) \equiv A-u(a) + \delta \quad \text{for} \quad a \leq t \leq a + \delta/2,$$

$$d^+(t) \equiv \delta \quad \text{for} \quad a+\delta \leq t \leq b,$$

$$d^-(t) \equiv u(b) - B+\delta \quad \text{for} \quad b - \delta/2 \leq t \leq b,$$

and

$$d^-(t) \equiv \delta \quad \text{for} \quad a \leq t \leq b-\delta.$$

Lastly, if $u(a) \geq A$ and $u(b) \leq B$, we define the domain

$$R^{\mp}(u) = \{(t,y) : a \leq t \leq b, \ -d^-(t) \leq y-u(t) \leq d^+(t)\},$$

where d^+, d^- are positive continuous functions satisfying

$$d^+(t) \equiv u(a) - A+\delta \quad \text{for} \quad a \leq t \leq a + \delta/2,$$

$$d^+(t) \equiv \delta \quad \text{for} \quad a+\delta \leq t \leq b,$$

$$d^-(t) \equiv B - u(b)+\delta \quad \text{for} \quad b - \delta/2 \leq t \leq b,$$

and

$$d^-(t) \equiv \delta \quad \text{for} \quad a \leq t \leq b-\delta.$$

If $u = u_s$ is a singular solution of (R), then the corresponding domains are denoted by $R^+(u_s)$, $R^-(u_s)$, $R^{\pm}(u_s)$ and $R^{\mp}(u_s)$.

As in the last chapter, we will also investigate paths formed by two or more intersecting solutions of the reduced problems. For the following four paths:

$$u_0(t) = \begin{cases} u_L(t), & a \leq t \leq t_0, \\ u_R(t), & t_0 \leq t \leq b, \end{cases}$$

$$u_1(t) = \begin{cases} u_L(t), & a \leq t \leq t_L, \\ u(t), & t_L \leq t \leq t_R, \\ u_R(t), & t_R \leq t \leq b, \end{cases}$$

$$u_2(t) = \begin{cases} u(t), & a \leq t \leq t_R, \\ u_R(t), & t_R \leq t \leq b, \end{cases}$$

and

$$u_3(t) = \begin{cases} u_L(t), & a \le t \le t_L, \\ u(t), & t_L \le t \le b, \end{cases}$$

we define the corresponding domains:

$$\mathcal{R}(u_0(t)) = \{(t,y) : a \le t \le b, \ |y-u_0(t)| \le \delta\},$$

$$\mathcal{R}(u_1(t)) = \{(t,y) : a \le t \le b, \ |y-u_1(t)| \le \delta\},$$

$$\mathcal{R}^+(u_2(t)) = \{(t,y) : a \le t \le b, \ -\delta \le y-u_2(t) \le d_2^+(t)\},$$

$$\mathcal{R}^-(u_2(t)) = \{(t,y) : a \le t \le b, \ -d_2^-(t) \le y - u_2(t) \le \delta\},$$

$$\mathcal{R}^+(u_3(t)) = \{(t,y) : a \le t \le b, \ -\delta \le y-u_3(t) \le d_3^+(t)\},$$

and

$$\mathcal{R}^-(u_3(t)) = \{(t,y) : a \le t \le b, \ -d_3^-(t) \le y-u_3(t) \le \delta\},$$

where d_2^+, d_2^-, d_3^+ and d_3^- are positive continuous functions satisfying, respectively,

$$d_2^+ \equiv A-u(a)+\delta \qquad \text{for} \quad a \le t \le a + \delta/2,$$

and

$$d_2^+(t) \equiv \delta \quad \text{for} \quad a+\delta \le t \le b, \text{ if } \ u(a) \le A;$$

$$d_2^-(t) \equiv u(a)-A+\delta \qquad \text{for} \quad a \le t \le a + \delta/2,$$

and

$$d_2^-(t) \equiv \delta \quad \text{for} \quad a+\delta \le t \le b, \quad \text{if} \quad u(a) \ge A;$$

$$d_3^+(t) \equiv B-u(b)+\delta \qquad \text{for} \quad b - \delta/2 \le t \le b,$$

and

$$d_3^+(t) \equiv \delta \quad \text{for} \quad a \le t \le b-\delta, \quad \text{if} \quad u(b) \le B;$$

$$d_3^-(t) \equiv u(b)-B+\delta \qquad \text{for} \quad b - \delta/2 \le t \le b,$$

and

$$d_3^-(t) \equiv \delta \quad \text{for} \quad a \le t \le b-\delta, \quad \text{if} \quad u(b) \ge B.$$

We now define the types of stability which solutions of the reduced problems (R_L), (R_R) and (R) are to possess in order to permit the study of the Dirichlet problem (DP_3) and other related boundary problems.

<u>Definition 5.1</u>. A solution $u = u_L(t)$ of the reduced problem (R_L) which exists in $[a,b]$ is said to be strongly (weakly) stable in $[a,b]$ if there exists a positive constant k such that

$$2p(t,y)u_L'(t) \geq k > 0 \quad (2p(t,y)u_L'(t) \geq 0) \quad \text{in} \quad R^+(u_L) \quad \text{or} \quad R^-(u_L).$$

Definition 5.2. A solution $u = u_R(t)$ of the reduced problem (R_R) which exists in $[a,b]$ is said to be strongly (weakly) stable in $[a,b]$ if there exists a positive constant k such that

$$2p(t,y)u_R'(t) \leq -k < 0 \quad (2p(t,y)u_R'(t) \leq 0) \quad \text{in} \quad R^+(u_R) \quad \text{or} \quad R^-(u_R).$$

Definition 5.3. A solution $u = u(t)$ of the reduced equation (R) is said to be (I_q)-, $(\tilde{I}_q)$-, (II_n)-, $(\tilde{II}_n)$-, (III_n)- or $(\tilde{III}_n)$-stable in $[a,b]$ if the function $h(t,y) = p(t,y)\,(u'(t))^2 + g(t,y)$ is (I_q)-,..., $(\tilde{III}_n)$-stable in the sense of Definitions 3.1 - 3.6, respectively.

Definition 5.4. A solution $u = u(t)$ of the reduced equation (R) is said to be locally strongly (weakly) stable if there exists a positive constant k and a small positive constant δ such that

$$2p(t,y)u'(t) \leq -k < 0 \quad (2p(t,y)u'(t) \leq 0) \quad \text{in} \quad R(u) \cap [a,a+\delta]$$

if $u(a) \neq A$, and

$$2p(t,y)u'(t) \geq k > 0 \quad (2p(t,y)u'(t) \geq 0) \quad \text{in} \quad R(u) \cap [b-\delta,b]$$

if $u(b) \neq B$ for $R(u) = R^+(u)$, $R^-(u)$, $R^{\pm}(u)$ or $R^{\mp}(u)$.

Definition 5.5. The reduced path $u = u_0(t)$ is said to be strongly (weakly) stable in $[a,b]$ if there exists a positive constant k (and a small positive constant δ) such that

$$2p(t,u_L(t))u_L'(t) \geq k > 0 \quad \text{in} \quad [a,t_0]$$

$$(2p(t,y)u_L'(t) \geq 0 \quad \text{in} \quad R(u_L,u_R) \cap [t_0-\delta,t_0])$$

and

$$2p(t,u_R(t))u_R'(t) \leq -k < 0 \quad \text{in} \quad [t_0,b]$$

$$(2p(t,y)u_R'(t) \leq 0 \quad \text{in} \quad R(u_L,u_R) \cap [t_0,t_0+\delta]).$$

Definition 5.6. The reduced path $u = u_1(t)$ is said to be strongly (weakly) stable in $[a,b]$ if there exists a positive constant k and a small positive constant δ such that

$$2p(t,u_L(t))u_L'(t) \geq k > 0 \quad \text{in} \quad [a,t_L]$$

$$(2p(t,y)u_L'(t) \geq 0 \quad \text{in} \quad R(u_L,u,u_R) \cap [t_L-\delta,t_L]),$$

$$2p(t,u(t))u'(t) \leq -k < 0 \quad \text{in} \quad [t_L,t_L+\delta]$$

$$(2p(t,y)u'(t) \leq 0 \quad \text{in} \quad R(u_L,u,u_R) \cap [t_L,t_L+\delta]),$$

$$2p(t,u(t))u'(t) \geq k > 0 \quad \text{in} \quad [t_R-\delta,t_R]$$

$$(2p(t,y)u'(t) \geq 0 \quad \text{in} \quad \mathcal{R}(u_L,u,u_R) \cap [t_R-\delta,t_R]),$$

and

$$2p(t,u_R(t))u_R'(t) \leq -k <0 \quad \text{in} \quad [t_R,b]$$

$$(2p(t,y)u_R'(t) \leq 0 \quad \text{in} \quad \mathcal{R}(u_L,u,u_R) \cap [t_R,t_R+\delta]).$$

Definition 5.7. The reduced path $u = u_2(t)$ is said to be strongly (weakly) stable in $[a,b]$ if there exists a positive constant k and a small positive constant δ such that

$$2p(t,y)u'(t) \leq -k < 0 \quad \text{in} \quad \mathcal{R}^{\pm}(u,u_R) \cap [a,a+\delta]$$

$$(2p(t,y)u'(t) \leq 0 \quad \text{in} \quad \mathcal{R}^{\pm}(u,u_R) \cap [a,a+\delta]),$$

$$2p(t,u(t))u'(t) \geq k > 0 \quad \text{in} \quad [t_R-\delta,t_R]$$

$$(2p(t,y)u'(t) \geq 0 \quad \text{in} \quad \mathcal{R}^{\pm}(u,u_R) \cap [t_R-\delta,t_R])$$

and

$$2p(t,u_R(t))u_R'(t) \leq -k < 0 \quad \text{in} \quad [t_R,b]$$

$$(2p(t,y)u_R'(t) \leq 0 \quad \text{in} \quad \mathcal{R}^{\pm}(u,u_R) \cap [t_R,t_R+\delta]).$$

Definition 5.8. The reduced path $u = u_3(t)$ is said to be strongly (weakly) stable in $[a,b]$ if there exists a positive constant k and a small positive constant δ such that

$$2p(t,u_L(t))u_L'(t) \geq k > 0 \quad \text{in} \quad [a,t_L]$$

$$(2p(t,y)u_L'(t) \geq 0 \quad \text{in} \quad \mathcal{R}(u_L,u) \cap [t_L-\delta,t_L]),$$

$$2p(t,u(t))u'(t) \leq -k < 0 \quad \text{in} \quad [t_L,t_L+\delta]$$

$$(2p(t,y)u'(t) \leq 0 \quad \text{in} \quad \mathcal{R}^{\pm}(u_L,u) \cap [t_L,t_L+\delta]),$$

$$2p(t,y)u'(t) \geq k > 0 \quad \text{in} \quad \mathcal{R}^{\pm}(u_L,u) \cap [b-\delta,b]$$

$$(2p(t,y)u'(t) \geq 0 \quad \text{in} \quad \mathcal{R}^{\pm}(u_L,u) \cap [b-\delta,b]).$$

With these definitions we are now in a position to give results on the types of asymptotic behavior displayed by solutions of the Dirichlet problem (DP_3). In what follows we tacitly assume that the functions p and g are continuous in t and sufficiently differentiable with respect to y in the domains defined above. Note that in the statement of each of the theorems below, the constant c is positive, depending only on the reduced solution under consideration.

Theorem 5.1. Assume that the reduced problem (R_L) or (R_R) has a strongly stable solution $u = u_L(t)$ or $u = u_R(t)$ of class $C^{(2)}([a,b])$. Assume also that there exists a positive constant ν such that either (i) $p(t,y) \geq \nu > 0$ in $\mathcal{R}^+(u_L)$ if $u_L(b) \leq B$ and $p(t,y) \leq -\nu < 0$ in $\mathcal{R}^-(u_L)$ if $u_L(b) \geq B$, or (ii) $p(t,y) \geq \nu > 0$ in $\mathcal{R}^+(u_R)$ if $u_R(a) \leq A$ and $p(t,y) \leq -\nu < 0$ in $\mathcal{R}^-(u_R)$ if $u_R(a) \geq A$.

Then there exists an $\varepsilon_0 > 0$ such that for $0 < \varepsilon \leq \varepsilon_0$ the problem (DP_3) has a solution $y = y(t,\varepsilon)$ in $[a,b]$ satisfying

$$-c\varepsilon \leq y(t,\varepsilon) - u_L(t) \leq w_R(t,\varepsilon) + c\varepsilon, \quad \text{if } u_L(b) \leq B,$$

$$-w_R(t,\varepsilon) - c\varepsilon \leq y(t,\varepsilon) - u_L(t) \leq c\varepsilon, \quad \text{if } u_L(b) \geq B,$$

or

$$-c\varepsilon \leq y(t,\varepsilon) - u_R(t) \leq w_L(t,\varepsilon) + c\varepsilon, \quad \text{if } u_R(a) \leq A,$$

$$-w_L(t,\varepsilon) - c\varepsilon \leq y(t,\varepsilon) - u_R(t) \leq c\varepsilon, \quad \text{if } u_R(a) \geq A,$$

where

$$w_R(t,\varepsilon) = -\varepsilon\nu^{-1}\ln\{(b-a)^{-1}[b-t+(t-a)\exp(-|B-u_L(b)|\nu\varepsilon^{-1})]\},$$

and

$$w_L(t,\varepsilon) = -\varepsilon\nu^{-1}\ln\{(b-a)^{-1}[t-a+(b-t)\exp(-|A-u_R(a)|\nu\varepsilon^{-1})]\}.$$

Proof: As the proofs of all cases are similar, we shall only give the proof for the case where the reduced equation (R_R) has a solution $u = u_R(t)$ such that $u_R(a) \leq A$ and $p(t,y) \geq \nu > 0$ in $\mathcal{R}^+(u_R)$.

We first linearize about u_R by setting $z = y - u_R$. This leads to

$$\varepsilon z'' = p(t,y)z'^2 + 2p(t,y)u_R'z' + g_y(t,\xi)z - \varepsilon u_R'',$$

$$z(a,\varepsilon) = A - u_R(a), \quad z(b,\varepsilon) = 0,$$

where (t,ξ) is some intermediate point between (t,u_R) and (t,u_R+z). Since $p(t,y) \geq \nu$ and $2pu_R' \leq -k < 0$ by definition of strong stability, and since g_y is bounded, say $|g_y| \leq \ell$ $(\ell > 0)$ in $\mathcal{R}^+(u_R)$, we are further led to the nonlinear differential equation

$$\varepsilon z'' = z'^2 - kz' - \ell z - \varepsilon u_R''.$$

The nonlinear (quadratic) part and the linear part will be utilized to construct the bounding pair of functions. Indeed the function

$$w_L(t,\varepsilon) = -(\varepsilon/\nu)\ln\{(b-a)^{-1}[t-a + (b-t)\exp(-|A-u_R(a)|\nu\varepsilon^{-1})]\}$$

is the solution of the nonlinear boundary value problem

$$\varepsilon w'' = \nu w'^2,$$

$$w(a,\varepsilon) = |A-u_R(a)|, \quad w(b,\varepsilon) = 0,$$

If $0 < \varepsilon < k^2/4\ell$, the characteristic equation $\varepsilon\lambda^2 + k\lambda + \ell = 0$ has two negative roots

$$\lambda = -\ell/k + O(\varepsilon) \quad \text{and} \quad \lambda_1 = -k/\varepsilon + O(\varepsilon).$$

Then, for any $\gamma > 0$, the function

$$\Gamma(t,\varepsilon,\gamma) = \varepsilon\gamma\ell^{-1}(\exp[-\lambda(b-t)] - 1)$$

is the solution of the linear non-homogeneous equation

$$\varepsilon\Gamma'' = -k\Gamma' - \ell\Gamma - \varepsilon\gamma$$

such that

$$\Gamma(b,\varepsilon,\gamma) = 0, \quad 0 \le \Gamma \le c\varepsilon \quad \text{and} \quad -c\varepsilon \le \Gamma' < 0,$$

for some $c > 0$.

We now define the bounding pair

$$\alpha(t,\varepsilon) = u_R(t) - \Gamma(t,\varepsilon,\gamma)$$

$$\beta(t,\varepsilon) = u_R(t) + w_L(t,\varepsilon) + \Gamma(t,\varepsilon,\tilde{\gamma}),$$

where γ and $\tilde{\gamma}$ are positive constants to be chosen so that the bounding pair satisfy the required inequalities of Theorem 2.1.

It follows from the definition that $\alpha \le \beta$, $\alpha(a,\varepsilon) \le A \le \beta(a,\varepsilon)$ and $\alpha(b,\varepsilon) \le B \le \beta(b,\varepsilon)$, since $u_R(b) = B$. Differentiating and applying Taylor's Theorem, we have

$$\begin{aligned}
\varepsilon\alpha'' - p(t,\alpha)\alpha'^2 - g(t,\alpha) &= \varepsilon u_R'' - \varepsilon\Gamma'' \\
&\quad - p(t,u_R-\Gamma)(u_R'-\Gamma')^2 - g(t,u_R-\Gamma) \\
&= \varepsilon u_R'' - \varepsilon\Gamma'' - 2pu_R'\Gamma' - p\Gamma'^2 \\
&\quad + g(t,u_R) - g(t,u_R-\Gamma) \\
&\ge -\varepsilon|u_R''| - (\varepsilon\Gamma'' + k\Gamma' + \ell\Gamma) - p\Gamma'^2 \\
&\ge \varepsilon(\gamma - |u_R''| - |p|c^2\varepsilon) > 0,
\end{aligned}$$

if we choose $\gamma = |u_R''| + 1$, for ε sufficiently small, say $0 < \varepsilon \le \varepsilon_1$.

Similarly,

$$\begin{aligned}\varepsilon\beta'' - p(t,\beta)\beta'^2 - g(t,\beta) &= \varepsilon u_R'' + \varepsilon w_L'' + \varepsilon\Gamma'' - pu_R'^2 - pw_L'^2 - p\tilde{\Gamma}'^2 \\ &\quad - 2pu_R'(w_L'+\tilde{\Gamma}') - 2pw_L'\tilde{\Gamma}' - g(t,\beta) \\ &\le \varepsilon u_R'' + \varepsilon w_L'' + \varepsilon\tilde{\Gamma}'' - \nu w_L'^2 + k\tilde{\Gamma}' + \ell(w_L+\tilde{\Gamma}), \\ &\le \varepsilon|u_R''| + \ell w_L - \varepsilon\tilde{\gamma} \\ &\le 0,\end{aligned}$$

if we choose ε sufficiently small, say $0 < \varepsilon \le \varepsilon_2$, so that $\ell w_L \le \varepsilon K$ for some $K > 0$, and then choose $\tilde{\gamma} = |u_R''| + K$.

Thus, for $0 < \varepsilon \le \varepsilon_0 = \min\{\varepsilon_1,\varepsilon_2\}$, the bounding pair satisfies the required inequalities of Theorem 2.1 and the conclusion follows.

A similar result holds if the functions u_L and u_R are only weakly stable in $[a,b]$, provided that some form of stability with respect to y is imposed. A precise discussion is contained in the following theorems.

Theorem 5.2. Assume that the reduced problem (R_L) or (R_R) has a weakly stable solution $u = u_L(t)$ or $u = u_R(t)$ of class $C^{(2)}([a,b])$ such that $p(t,y)(B-u_L(b)) \ge 0$ in $\mathcal{R}^{\pm}(u_L) \cap [b-\delta,b]$ or $p(t,y)(A-u_R(a)) \ge 0$ in $\mathcal{R}^{\pm}(u_R) \cap [a,a+\delta]$. Assume also that u_L or u_R is (I_q)- or $(\tilde{I}_q)$-stable in $[a,b]$. Then there exists an $\varepsilon_0 > 0$ such that for $0 < \varepsilon \le \varepsilon_0$, the problem (DP_3) has a solution $y = y(t,\varepsilon)$ in $[a,b]$ satisfying

$$-c\varepsilon^{1/(2q+1)} \le y(t,\varepsilon) - u_L(t) \le w_R(t,\varepsilon) + c\varepsilon^{1/(2q+1)} \quad \text{if} \quad u_L(b) \le B,$$

$$-w_R(t,\varepsilon) - c\varepsilon^{1/(2q+1)} \le y(t,\varepsilon) - u_L(t) \le c\varepsilon^{1/(2q+1)} \quad \text{if} \quad u_L(b) \ge B,$$

or

$$-c\varepsilon^{1/(2q+1)} \le y(t,\varepsilon) - u_R(t) \le w_L(t,\varepsilon) + c\varepsilon^{1/(2q+1)} \quad \text{if} \quad u_R(a) \le A,$$

$$-w_L(t,\varepsilon) - c\varepsilon^{1/(2q+1)} \le y(t,\varepsilon) - u_R(t) \le c\varepsilon^{1/(2q+1)} \quad \text{if} \quad u_R(a) \ge A,$$

where the functions w_L and w_R are as defined in the conclusion of Theorem 3.1 with u replaced by u_R and u_L, respectively.

Proof: Let us give the proof for the case when (R_L) has an (I_0)-stable solution $u = u_L(t)$ such that $u_L(b) \ge B$ and $p \le 0$ in $\mathcal{R}^-(u_L)$ for $b-\delta \le t \le b$. The proofs for the other cases are analogous.

As noted in the proof of Theorem 3.1, the function

$$w_R(t,\varepsilon) = |B-u_L(b)|\exp[-(m\varepsilon^{-1})^{1/2}(b-t)]$$

is the solution of

$$\varepsilon w'' = mw,$$

$$w(b,\varepsilon) = |B - u_L(b)| = u_L(b)-B, \quad w'(b,\varepsilon) = (m\varepsilon^{-1})^{1/2}|B-u_L(b)|.$$

Let us define the bounding pair

$$\alpha(t,\varepsilon) = u_L(t) - w_R(t,\varepsilon) - \varepsilon\gamma m^{-1},$$

$$\beta(t,\varepsilon) = u_L(t) + \varepsilon M m^{-1},$$

where $M = |u_L''|$, and γ is a positive constant to be chosen later.

Since β clearly satisfies all the required inequalities of Theorem 2.1, we turn our attention to α. Differentiating and applying Taylor's Theorem (with (t,ξ) as the intermediate point), we obtain

$$\begin{aligned}\varepsilon\alpha'' - p(t,\alpha)\alpha'^2 - g(t,\alpha) &= \varepsilon u_L'' - \varepsilon w_R'' - p(t,\alpha)u_L'^2 - p(t,\alpha)w_R'^2 \\ &\quad + 2p(t,\alpha)u_L'w_R' - g(t,\alpha) \\ &\geq -\varepsilon M - mw_R - pw_R'^2 + 2pu_L'w_R' \\ &\quad + [p_y(t,\xi)u_L'^2 + g_y(t,\xi)](w_R + \varepsilon\gamma m^{-1}) \\ &\geq -\varepsilon M - pw_R'^2 + \varepsilon\gamma,\end{aligned}$$

since $pu_L'w_R' \geq 0$ by the weak stability of u_L, and $p_y u_L'^2 + g_y \geq m > 0$ by the (I_0)-stability of u_L. Now, in the interval $[b-\delta/2,b]$, we have $p \leq 0$ and so the desired inequality follows by setting $\gamma \geq M$. In the remaining interval $[a,b-\delta/2)$, there exist $c_1 > 0$ and $\varepsilon_0 > 0$ such that $pw_R'^2 \leq c_1\varepsilon$ for $0 < \varepsilon \leq \varepsilon_0$; and so the desired inequality follows if we set $\gamma \geq M + c_1$. Theorem 5.2 now follows from Theorem 2.1.

A similar argument allows us to prove the following theorem.

<u>Theorem 5.3</u>. Assume that the reduced problem (R_L) or (R_R) has a weakly stable solution $u = u_L(t)$ or $u = u_R(t)$ of class $C^{(2)}([a,b])$ which is also (II_n)- or $(\tilde{II}_n)$-stable in $[a,b]$. Assume also that $u_L'' \geq 0$ or $u_R'' \geq 0$ in (a,b), $u_L(b) \leq B$ or $u_R(a) \leq A$ and $p(t,y) \geq 0$ in $R^+(u_L) \cap [b-\delta,b]$ or $p(t,y) \geq 0$ in $R^+(u_R) \cap [a,a+\delta]$. Then there exists an $\varepsilon_0 > 0$ such that for $0 < \varepsilon \leq \varepsilon_0$ the problem (DP_3) has a solution $y = y(t,\varepsilon)$ in $[a,b]$ satisfying

$$0 \leq y(t,\varepsilon) - u_L(t) \leq w_R(t,\varepsilon) + c\varepsilon^{1/n}$$

or

$$0 \le y(t,\varepsilon) - u_R(t) \le w_L(t,\varepsilon) + c\varepsilon^{1/n},$$

where the functions w_L and w_R are as defined in the conclusion of Theorem 3.2 with u replaced by u_R and u_L, respectively.

The next result follows from Theorem 5.3 with the change of variable $y \to -y$.

Theorem 5.4. Assume that the reduced problem (R_L) or (R_R) has a weakly stable solution $u = u_L(t)$ or $u = u_R(t)$ of class $C^{(2)}([a,b])$ which is also (III_n)- or $(\tilde{III}_n)$-stable in $[a,b]$. Assume also that $u_L'' \le 0$ or $u_R'' \le 0$ in (a,b), $u_L(b) \ge B$ or $u_R(a) \ge A$, and $p(t,y) \le 0$ in $\mathcal{R}^-(u_L) \cap [b-\delta,b]$ or $p(t,y) \le 0$ in $\mathcal{R}^-(u_R) \cap [a,a+\delta]$. Then there exists an $\varepsilon_0 > 0$ such that for $0 < \varepsilon \le \varepsilon_0$ the problem (DP_3) has a solution $y = y(t,\varepsilon)$ in $[a,b]$ satisfying

$$-w_R(t,\varepsilon) - c\varepsilon^{1/n} \le y(t,\varepsilon) - u_L(t) \le 0$$

or

$$-w_L(t,\varepsilon) - c\varepsilon^{1/n} \le y(t,\varepsilon) - u_R(t) \le 0,$$

where the functions w_L and w_R are as defined in the conclusion of Theorem 3.2 with u replaced by u_R and u_L, respectively.

In the next two theorems, the reduced equation (R) has a solution $u = u(t)$ which in general satisfies neither of the boundary conditions but which is locally strongly or weakly stable. The proofs are similar to those of Theorems 5.1 and 5.2 and are omitted.

Theorem 5.5. Assume that the reduced equation (R) has a locally strongly (weakly) stable solution $u = u(t)$ of class $C^{(2)}([a,b])$ such that $p(t,y) \ge \nu > 0$ $(p(t,y) \ge 0)$ in $\mathcal{R}^+(u) \cap \{[a,a+\delta] \cup [b-\delta,b]\}$ if $u(a) \le A$ and $u(b) \le B$, $p(t,y) \le -\nu < 0$ $(p(t,y) \le 0)$ in $\mathcal{R}^-(u) \cap \{[a,a+\delta] \cup [b-\delta,b]\}$ if $u(a) \ge A$ and $u(b) \ge B$, $p(t,y) \ge \nu > 0$ $(p(t,y) \ge 0)$ in $\mathcal{R}^\pm(u) \cap [a,a+\delta]$ and $p(t,y) \le -\nu < 0$ $(p(t,y) \le 0)$ in $\mathcal{R}^\pm(u) \cap [b-\delta,b]$ if $u(a) \le A$ and $u(b) \ge B$, and $p(t,y) \le -\nu < 0$ $(p(t,y) \le 0)$ in $\mathcal{R}^+(u) \cap [a,a+\delta]$ and $p(t,y) \ge \nu > 0$ $(p(t,y) > 0)$ in $\mathcal{R}^+(u) \cap [b-\delta,b]$ if $u(a) \ge A$ and $u(b) \le B$ for a positive constant ν. Assume also that u is (I_q)- or $(\tilde{I}_q)$-stable in $[a,b]$. Then there exists an $\varepsilon_0 > 0$ such that for $0 < \varepsilon \le \varepsilon_0$ the problem (DP_3) has a solution $y = y(t,\varepsilon)$ in $[a,b]$ satisfying

$$-c\varepsilon^{1/(2q+1)} \le y(t,\varepsilon) - u(t) \le w_L(t,\varepsilon) + w_R(t,\varepsilon) + c\varepsilon^{1/(2q+1)}$$

if $u(a) \le A$ and $u(b) \le B$;

$$-w_L(t,\varepsilon) - c\varepsilon^{1/(2q+1)} \leq y(t,\varepsilon) - u(t) \leq w_R(t,\varepsilon) + c\varepsilon^{1/(2q+1)}$$

if $u(a) \geq A$ and $u(b) \leq B$;

$$-w_R(t,\varepsilon) - c\varepsilon^{1/(2q+1)} \leq y(t,\varepsilon) - u(t) \leq w_L(t,\varepsilon) + c\varepsilon^{1/(2q+1)}$$

if $u(a) \leq A$ and $u(b) \geq B$; or

$$-w_L(t,\varepsilon) - w_R(t,\varepsilon) - c\varepsilon^{1/[2q+1)} \leq y(t,\varepsilon) - u(t) \leq c\varepsilon^{1/(2q+1)}$$

if $u(a) \geq A$ and $u(b) \geq B$. Here

$$w_L(t,\varepsilon) = -\varepsilon\nu^{-1}\ln\{(b-a)^{-1}[t-a + (b-t)\exp(-|A-u(a)|\nu\varepsilon^{-1})]\}$$

and

$$w_R(t,\varepsilon) = -\varepsilon\nu^{-1}\ln\{(b-a)^{-1}[b-t + (t-a)\exp(-|B-u(b)|\nu\varepsilon^{-1})]\}$$

if u is locally strongly stable, and w_L, w_R are as defined in the conclusion of Theorem 3.1 if u is locally weakly stable.

Theorem 5.6. Assume that the reduced equation (R) has a locally strongly (weakly) stable solution $u = u(t)$ of class $C^{(2)}([a,b])$ which is also (II_n)- or $(\tilde{II}_n)$-stable in $[a,b]$. Assume also that $u'' \geq 0$ in (a,b), $u(a) \leq A$, $u(b) \leq B$, $p(t,y) \geq \nu > 0$ $(p(t,y) \geq 0)$ in $\mathcal{R}^+(u) \cap \{[a,a+\delta] \cup [b-\delta,b]\}$ for a positive constant ν. Then there exists an $\varepsilon_0 > 0$ such that for $0 < \varepsilon \leq \varepsilon_0$ the problem (DP_3) has a solution $y = y(t,\varepsilon)$ in $[a,b]$ satisfying

$$0 \leq y(t,\varepsilon) - u(t) \leq w_L(t,\varepsilon) + w_R(t,\varepsilon) + c\varepsilon^{1/n},$$

where w_L, w_R are as defined in Theorem 5.5 (Theorem 3.2) for the locally strongly (weakly) stable solution.

An analogous result holds when the locally strongly or weakly stable function u is (III_n)- or $(\tilde{III}_n)$-stable, provided that $u'' \leq 0$ in (a,b), $u(a) \geq A$, $u(b) \geq B$ and $p(t,y) \leq -\nu < 0$ or $p(t,y) \leq 0$ in $\mathcal{R}^-(u) \cap \{[a,a+\delta] \cup [b-\delta,b]\}$. It can be proved by making the change of variable $y \to -y$ and applying Theorem 5.6 to the transformed problem.

We remark that if the solution u of (R) is a singular solution $u = u_s \equiv$ const., then $2p(t,y)u_s'(t) \equiv 0$ in $[a,b]$ and so u_s is certainly locally weakly stable. Consequently Theorems 5.5 and 5.6 apply, if u_s is q- or n-stable and satisfies the additional geometric conditions of these theorems.

§5.3. Robin Problems: Boundary Layer Phenomena

We turn now to a consideration of similar phenomena for the Robin problems

$$\varepsilon y'' = p(t,y)y'^2 + g(t,y), \quad a < t < b,$$
$$y(a,\varepsilon) - p_1 y'(a,\varepsilon) = A, \quad y(b,\varepsilon) = B, \tag{RP_5}$$

and

$$\varepsilon y'' = p(t,y)y'^2 + g(t,y), \quad a < t < b,$$
$$y(a,\varepsilon) - p_1 y'(a,\varepsilon) = A, \quad y(b,\varepsilon) + p_2 y'(b,\varepsilon) = B, \tag{RP_6}$$

where p_1 and p_2 are positive constants. (The related problem $\varepsilon y'' = p(t,y)y'^2 + g(t,y)$, $a < t < b$, $y(a,\varepsilon)$, $y(b,\varepsilon) + p_2 y'(b,\varepsilon)$ prescribed, can be studied by making the change of variable $t \to a+b-t$ and applying the theory of (RP_5) to the transformed problem.) The associated reduced problems are then

$$p(t,u)u'^2 + g(t,u) = 0, \quad a < t < t_1 \leq b,$$
$$u(a) - p_1 u'(a) = A, \tag{$\tilde{R}_L$}$$

$$p(t,u)u'^2 + g(t,u) = 0, \quad a \leq t_2 < t < b,$$
$$u(b) = B, \tag{R_R}$$

$$p(t,u)u'^2 + g(t,u) = 0, \quad a \leq t_2 < t < b,$$
$$u(b) + p_2 u'(b) = B, \tag{$\tilde{R}_R$}$$

and

$$p(t,u)u'^2 + g(t,u) = 0, \quad a < t < b, \tag{R}$$

whose solutions are denoted by u_L, u_R, and u, respectively. The definitions of stability given at the beginning of this chapter are assumed to apply to the functions u_L, u_R and u with the following modifications. In the case of (RP_5) the domains $\mathcal{R}^+(u_R)$ and $\mathcal{R}^-(u_R)$ are replaced by

$$\mathcal{R}(u_R) = \{(t,y)\colon a \leq t \leq b, \ |y - u_R(t)| \leq \delta\},$$

while in the definition of the domains $\mathcal{R}^+(u)$, $\mathcal{R}^-(u)$, $\mathcal{R}^{\pm}(u)$ and $\mathcal{R}^{\mp}(u)$ the error function is assumed to be uniformly small (that is, bounded above by δ) in $[a,b-\delta]$. Similarly, in the case of (RP_6) the error function in each of the domains is assumed to be uniformly small in $[a,b]$.

The theory for the problems (RP_5) and (RP_6) is not as straightforward as for their counterparts in Chapter IV, in view of the nonlinear (quadratic) dependence on y'.

We shall discuss the Robin problem (RP_5) first. The following two results relate to the solution u_L of the reduced problem $(\tilde{R}_L)$.

Theorem 5.7. Assume that the reduced problem $(\tilde{R}_L)$ has a strongly (weakly) stable solution $u = u_L(t)$ of class $C^{(2)}([a,b])$ which is also (I_q)- or $(\tilde{I}_q)$-stable in $[a,b]$. Assume also that there exists a positive constant ν such that $p(t,y) \geq \nu > 0$ $(p(t,y) \geq 0)$ in $R^+(u_L)$ if $u_L(b) \leq B$ and $p(t,y) \leq -\nu < 0$ $(p(t,y) \leq 0)$ in $R^-(u_L)$ if $u_L(b) \geq B$. Then there exists an $\varepsilon_0 > 0$ such that for $0 < \varepsilon \leq \varepsilon_0$, the problem (RP_5) has a solution $y = y(t,\varepsilon)$ in $[a,b]$ satisfying

$$-c\varepsilon^{1/(2q+1)} \leq y(t,\varepsilon) - u_L(t) \leq w_R(t,\varepsilon) + c\varepsilon^{1/(2q+1)}, \quad \text{if } u_L(b) \leq B,$$

and

$$-w_R(t,\varepsilon) - c\varepsilon^{1/(2q+1)} \leq y(t,\varepsilon) - u_L(t) \leq c\varepsilon^{1/(2q+1)}, \quad \text{if } u_L(b) \geq B,$$

where w_L and w_R are as defined in Theorem 5.1 (Theorem 3.1) with u replaced by u_L, for the strongly (weakly) stable solution.

Proof: We only give the proof for the case when $(\tilde{R}_L)$ has a strongly stable solution $u = u_L(t)$, which is also (I_0)-stable and satisfies $u_L(b) \leq B$, with $p \geq \nu > 0$ in $R^+(u_L)$.

Define the functions

$$\alpha(t,\varepsilon) = u_L(t) - \varepsilon M m^{-1},$$

$$\beta(t,\varepsilon) = u_L(t) + w_R(t,\varepsilon) + \varepsilon M m^{-1},$$

where $M \geq |u_L''|$. It is straightforward to show that $\alpha \leq \beta$, $\alpha(a,\varepsilon) - p_1\alpha'(a,\varepsilon) \leq A \leq \beta(a,\varepsilon) - p_1\beta'(a,\varepsilon)$, and $\alpha(b,\varepsilon) \leq B \leq \beta(b,\varepsilon)$; moreover, α,β satisfy the required differential inequalities (cf. the proofs of Theorem 5.1 and 5.2). Thus the conclusion of Theorem 5.7 follows from Theorem 2.3.

Similarly the following result can be proved by applying Theorem 2.3.

Theorem 5.8. Assume that the reduced problem $(\tilde{R}_L)$ has a strongly (or weakly) stable solution $u = u_L(t)$ of class $C^{(2)}([a,b])$ which is also (II_n)- or $(\tilde{II}_n)$-stable in $[a,b]$. Assume also that $u_L'' \geq 0$ in (a,b), $u_L(b) \leq B$ and $p(t,y) \geq \nu > 0$ (or $p(t,y) \geq 0$) in $R^+(u_L)$. Then there exists an $\varepsilon_0 > 0$ such that for $0 < \varepsilon \leq \varepsilon_0$, the problem (RP_5) has a solution $y = y(t,\varepsilon)$ in $[a,b]$ satisfying

$$0 \leq y(t,\varepsilon) - u_L(t) \leq w_R(t,\varepsilon) + c\varepsilon^{1/n},$$

where w_R is as defined in Theorem 5.1 (or Theorem 3.2 with u_L replacing u) for the strongly (or weakly) stable solution u_L.

A corresponding result holds for (III_n)- or $(\tilde{III}_n)$-stable solutions u_L, provided that $u_L'' \leq 0$ in (a,b), $u_L(b) \geq B$ and $p(t,y) \leq -\nu < 0$ (or $p(t,y) \leq 0$) in $\mathcal{R}^-(u_L)$. We leave the precise formulation of this result to the reader.

Let us now study the case when the reduced problem (R_R) has a strongly stable solution $u = u_R(t)$. We expect that if u_R satisfies an appropriate inequality at $t = a$, then the boundary value problem (RP_5) has a solution $y = y(t,\varepsilon)$ satisfying

$$\lim_{\varepsilon \to 0^+} y(t,\varepsilon) = u_R(t) \quad \text{in} \quad [a,b].$$

More precisely, we have the following result.

Theorem 5.9. Assume that the reduced problem (R_R) has a solution $u = u_R(t)$ of class $C^{(2)}([a,b])$ which is strongly stable and (I_0)-stable. Let $L = \max_{\mathcal{R}(u_R)} |p(t,y)|$. Then, if

$$k_1 = k - Lp_1^{-1}|u_R(a) - p_1 u_R'(a) - A| > 0, \tag{$\dagger$}$$

there exists an $\varepsilon_0 > 0$ such that for $0 < \varepsilon \leq \varepsilon_0$, the problem (RP_5) has a solution $y = y(t,\varepsilon)$ in $[a,b]$ satisfying

$$|y(t,\varepsilon) - u_R(t)| \leq v_L(t,\varepsilon) + \varepsilon M m^{-1},$$

where

$$v_L(t,\varepsilon) = \varepsilon(p_1 k_1)^{-1}|u_R(a) - p_1 u_R'(a) - A| \exp[-k_1(t-a)\varepsilon^{-1}].$$

Note: The inequality in $(\dagger)$ can be motivated by considering, in place of (RP_5), the initial value problem

$$\begin{aligned} &\varepsilon y'' = p(t,y)y'^2 + g(t,y), \quad a < t < a+\delta, \\ &y(a,\varepsilon) = u_R(a), \quad y'(a,\varepsilon) = p_1^{-1}[u_R(a) - A]. \end{aligned} \tag{I}$$

It follows from the study of initial value problems (cf. [61], [93; Chapter 10], [87; Chapter 1]) that the solution $y = y(t,\varepsilon)$ of (I) will converge to u_R as $\varepsilon \to 0^+$, provided the "initial jump" $|y'(a,\varepsilon) - u_R'(a)|$ is sufficiently small, that is, provided $|u_R(a) - p_1 u_R'(a) - A| < kp_1/L$.

Proof: We prove only the case for which $p(t,y) > 0$, since the proof for the case $p(t,y) < 0$ is similar. Linearizing about u_R by setting $z = y - u_R$, we are led to the Robin problem

$$\varepsilon z'' = p(t,y)(z'+2u_R')z' + [p_y(t,\xi)u_R'^2 + g_y(t,\xi)]z - \varepsilon u_R''$$
$$z(a) - p_1 z'(a) = A - u_R(a) + p_1 u_R'(a) = \gamma$$
$$z(b) = 0,$$

where (t,ξ) is some intermediate point between (t,u_R) and (t,u_R+z). Since $2pu_R' \le -k$ and $p_y u_R'^2 + g_y \ge m > 0$ by assumption, we are further led to the equation

$$\varepsilon z'' = [p(t,y)z' - k]z' + mz - \varepsilon u_R''.$$

Indeed, we note that the positive function $v = v_L(t,\varepsilon)$ is the solution of $\varepsilon v'' = -k_1 v'$ with the properties:

$$v_L = O(\varepsilon), \quad v_L'(a) = -\gamma/p_1 \quad \text{and} \quad 0 > p(t,y)v_L' > k_1-k.$$

Suppose first that the reduced problem u_R is such that $u_R(a)$ $p_1 u_R'(a) > A$. We then define the following pair of functions:

$$\alpha(t,\varepsilon) = u_R(t) - v_L(t,\varepsilon) - \varepsilon M m^{-1},$$
$$\beta(t,\varepsilon) = u_R(t) + \varepsilon M m^{-1},$$

where $M \ge |u_R''|$. The function β clearly satisfies the required inequalities. It is easy to verify that the function α satisfies $\alpha(a,\varepsilon) - p_1\alpha'(a,\varepsilon) < A$ and $\alpha(b,\varepsilon) > B$; moreover,

$$\begin{aligned}\varepsilon\alpha'' - p(t,\alpha)\alpha'^2 - g(t,\alpha) &= \varepsilon u_R'' - \varepsilon v_L'' + [p(t,u_R)-p(t,\alpha)]u_R'^2 + 2pu_R'v_L' \\ &\quad - pv_L'^2 + g(t,u_R) - g(t,\alpha) \\ &\ge -\varepsilon M + k_1 v_L' + m(v_L + \varepsilon M m^{-1}) - kv_L' - pv_L'^2 \\ &= mv_L + (k_1-k-pv_L')v_L' > 0,\end{aligned}$$

by definition of k_1 and v_L.

Suppose finally that $u_R(a) - p_1 u_R'(a) \le A$. In this case, we define the following pair of functions

$$\alpha(t,\varepsilon) = u_R(t) - \varepsilon M m^{-1},$$
$$\beta(t,\varepsilon) = u_R(t) + v_L(t,\varepsilon) + \varepsilon M m^{-1},$$

with $k_1 = k$ in the definition of v_L. The function α clearly satisfies the required inequalities. It is easy to verify that the function β satisfies $\beta(a,\varepsilon) - p_1\beta'(a,\varepsilon) \geq A$ and $\beta(b,\varepsilon) \geq B$; moreover,

$$\begin{aligned} p(t,\beta)\beta'^2 + g(t,\beta) - \varepsilon\beta'' &= [p(t,\beta)-p(t,u_R)]u_R'^2 + 2pu_R'v_L' + pv_L'^2 \\ &\quad + g(t,\beta) - g(t,u_R) - \varepsilon v_L'' - \varepsilon u_R'' \\ &\geq m(v_L + Mm^{-1}) - kv_L' + k_1v_L' - \varepsilon M > 0. \end{aligned}$$

Theorem 5.9 now follows from Theorem 2.3.

Such a strong result is not possible if the reduced solution u_R of (R_R) is not strongly stable, but only weakly stable. However, if we impose an additional assumption, we can obtain the following two theorems. Only the first theorem will be proved, since the proof for the second theorem is similar.

Theorem 5.10. Assume that the reduced problem (R_R) has a solution $u = u_R(t)$ of class $C^{(2)}([a,b])$ which is weakly stable and (I_q)-stable. Then, if

$$p(t,y)[A - u_R(a) + p_1u_R'(a)] \geq 0$$

in $R(u_R) \cap [a,a+\delta]$, there exists an $\varepsilon_0 > 0$ such that for $0 < \varepsilon \leq \varepsilon_0$, the problem (RP_5) has a solution $y = y(t,\varepsilon)$ in $[a,b]$ satisfying

$$|y(t,\varepsilon) - u_R(t)| \leq v_L(t,\varepsilon) + c\varepsilon^{1/(2q+1)},$$

where v_L is given in Theorem 3.4 with u replaced by u_R.

Proof: We give the proof only for the case that u_R is (I_0)-stable and such that $u_R(a) - p_1u_R'(a) \leq A$, and $p \geq 0$ in $R(u_R)$ for $a \leq t \leq a+\delta$. The other cases can be proved in a similar manner.

Define, for $a \leq t \leq b$ and $\varepsilon > 0$, the functions

$$\alpha(t,\varepsilon) = u_R(t) - \varepsilon Mm^{-1},$$

$$\beta(t,\varepsilon) = u_R(t) + v_L(t,\varepsilon) + \varepsilon(M+1)m^{-1},$$

where $M \geq |u_R''|$, and

$$v_L(t,\varepsilon) = (\varepsilon/mp_1^2)^{1/2}|u_R(a)-p_1u_R'(a)-A|\exp[-(m/\varepsilon)^{1/2}(t-a)]$$

is the solution of $\varepsilon v_L'' = mv_L$, such that $v_L'(a,\varepsilon) = \frac{1}{p_1}(u_R(a)-p_1u_R'(a)-A)$. Then, as in the proof of the previous theorem, these functions can be shown to satisfy the required inequalities.

Theorem 5.11. Assume that the reduced problem (R_R) has a solution $u = u_R(t)$ of class $C^{(2)}([a,b])$ which is weakly stable and (II_n)-stable, and $u_R'' \geq 0$. Then, if

$$u_R(a) - p_1 u_R'(a) \leq A \quad \text{and} \quad p(t,y) \geq 0$$

in $R(u_R) \cap [a,a+\delta]$, there exists an $\varepsilon_0 > 0$ such that for $0 < \varepsilon \leq \varepsilon_0$, the problem (RP_5) has a solution $y = y(t,\varepsilon)$ in $[a,b]$ satisfying

$$0 \leq y(t,\varepsilon) - u_R(t) \leq v_L(t,\varepsilon) + c\varepsilon^{1/n},$$

where v_L is given in Theorem 3.5 with u replaced by u_R.

A result analogous to Theorem 5.11 can be obtained if the reduced solution u_R is weakly stable and (III_n)-stable, and we leave it to the reader to provide details.

Similar results can be obtained, if we make similar assumptions with respect to the solution $u = u(t)$ of the reduced equation (R) which, in general, does not satisfy any of the given boundary conditions. We state only two representative results, and we omit proofs, since these are combinations of the proofs of Theorems 5.7, 5.8 and 5.9.

Theorem 5.12. Assume that the reduced equation (R) has a solution $u = u(t)$ of class $C^{(2)}([a,b])$ which is locally strongly stable and also (I_q)-stable or $(\tilde{I}_q)$-stable. Assume also that $p(t,y)[B-u(b)] \geq 0$ in $R^+(u) \cap [b-\delta,b]$ or in $R^-(u) \cap [b-\delta,b]$. Then, if

$$k_1 = k - Lp_1^{-1}|u_R(a) - p_1 u_R'(a) - A| > 0,$$

there exists an $\varepsilon_0 > 0$ such that for $0 < \varepsilon \leq \varepsilon_0$, the problem (RP_5) has a solution $y = y(t,\varepsilon)$ in $[a,b]$ satisfying

(a) $-c\varepsilon^{1/(2q+1)} \leq y(t,\varepsilon) - u(t) \leq v_L(t,\varepsilon) + w_R(t,\varepsilon) + c\varepsilon^{1/(2q+1)}$,

if $u(a) - p_1 u'(a) \leq A$ and $u(b) \leq B$;

(b) $-v_L(t,\varepsilon) - c\varepsilon^{1/(2q+1)} \leq y(t,\varepsilon) - u(t) \leq w_R(t,\varepsilon) + c\varepsilon^{1/(2q+1)}$,

if $u(a) - p_1 u'(a) \geq A$ and $u(b) \leq B$;

(c) $-w_R(t,\varepsilon) - c\varepsilon^{1/(2q+1)} \leq y(t,\varepsilon) - u(t) \leq v_L(t,\varepsilon) + c\varepsilon^{1/(2q+1)}$,

if $u(a) - p_1 u'(a) \leq A$ and $u(b) \geq B$; and

(d) $-v_L(t,\varepsilon) - w_R(t,\varepsilon) - c\varepsilon^{1/(2q+1)} \leq y(t,\varepsilon) - u(t) \leq c\varepsilon^{1/(2q+1)}$,

if $u(a) - p_1 u'(a) \geq A$ and $u(b) \geq B$.

Here, v_L is given in Theorem 5.9 and w_R is given in Theorem 5.1, with u_L replaced by u.

Theorem 5.13. Assume that the reduced equation (R) has a solution $u = u(t)$ of class $C^{(2)}([a,b])$ which is locally weakly stable and also (I_q)-stable or $(\tilde{I}_q)$-stable. Assume also that

$$p(t,y)[A - u(a) + p_1u'(a)] \geq 0$$

in $\mathcal{R}^+(u) \cap [a,a+\delta]$ or in $\mathcal{R}^-(u) \cap [a,a+\delta]$, and that

$$p(t,y)[B - u(b)] \geq 0$$

in $\mathcal{R}^+(u) \cap [b-\delta,b]$ or in $\mathcal{R}^-(u) \cap [b-\delta,b]$. Then there exists an $\varepsilon_0 > 0$ such that for $0 < \varepsilon \leq \varepsilon_0$, the problem (RP_5) has a solution $y = y(t,\varepsilon)$ in $[a,b]$ satisfying

(a) $-c\varepsilon^{1/(2q+1)} \leq y(t,\varepsilon) - u(t) \leq v_L + w_R + c\varepsilon^{1/(2q+1)}$,

if $u(a) - p_1u'(a) \leq A$ and $u(b) \leq B$;

(b) $-v_L - c\varepsilon^{1/(2q+1)} \leq y(t,\varepsilon) - u(t) \leq w_R + c\varepsilon^{1/(2q+1)}$,

if $u(a) - p_1u'(a) \geq A$ and $u(b) \leq B$;

(c) $-w_R - c\varepsilon^{1/(2q+1)} \leq y(t,\varepsilon) - u(t) \leq v_L + c\varepsilon^{1/(2q+1)}$,

if $u(a) - p_1u'(a) \leq A$ and $u(b) \geq B$; and

(d) $-v_L - w_R - c\varepsilon^{1/(2q+1)} \leq y(t,\varepsilon) - u(t) \leq c\varepsilon^{1/(2q+1)}$,

if $u(a) - p_1u'(a) \geq A$ and $u(b) \geq B$.

Here, v_L is given in Theorem 3.4 and w_R is given in Theorem 3.1.

We expect analogous results for the Robin problem (RP_6) as well. We state only the following result, leaving it to the reader to formulate others.

Theorem 5.14. Assume that the reduced problem $(\tilde{R}_L)$ (or $(\tilde{R}_R)$) has a solution $u = u(t)$ of class $C^{(2)}([a,b])$ which is strongly stable and also (I_0)-stable. Let $L = \max_{\mathcal{R}(u)} |p(t,y)|$. Then if

$$k_2 = k - Lp_2^{-1}|u_L(b) + p_2u_L'(b) - B| < 0,$$

(or if $k_1 = k - Lp_1^{-1}|u_R(a) - p_1u_R'(a) - A| > 0$), there exists an $\varepsilon_0 > 0$ such that for $0 < \varepsilon \leq \varepsilon_0$, the problem (RP_6) has a solution $y = y(t,\varepsilon)$ in $[a,b]$ satisfying

$$|y(t,\varepsilon) - u_L(t)| \leq v_R(t,\varepsilon) + \varepsilon Mm^{-1},$$

(or

$$|y(t,\varepsilon) - u_R(t)| \leq v_L(t,\varepsilon) + \varepsilon Mm^{-1}).$$

Here

$$v_R(t,\varepsilon) = \varepsilon(p_2k_2)^{-1}|u_L(b) + p_2u_L'(b) - B|\exp[-k_2(b-t)\varepsilon^{-1}]$$

and $v_L(t,\varepsilon)$ is given in Theorem 5.9.

§5.4. Interior Layer Phenomena

We turn now to the study of the interior crossing phenomena exhibited by the solutions of the problems (DP_3), (RP_5) and (RP_6). The theory for these problems differs from that discussed in Chapter IV because of two main factors. Firstly, as we have remarked earlier, the reduced equation

$$p(t,u)u'^2 + g(t,u) = 0 \tag{R}$$

may have singular solutions u_s = constant, which are the envelopes of the one-parameter family of solutions of (R). Secondly, since (R) is a quadratic function of u', there are two one-parameter families of solutions which can intersect one another within $[a,b]$ with *unequal* slopes. This situation can be seen from the reduced equation of Example (E_{12}), namely,

$$1 - u'^2 = 0, \tag{R_{12}}$$

whose solutions are $u = t + c$ and $u = -t + c$.

Let us first discuss the Dirichlet problem

$$\begin{aligned} &\varepsilon y'' = p(t,y)y'^2 + g(t,y), \quad a < t < b, \\ &y(a,\varepsilon) = A, \quad y(b,\varepsilon) = B. \end{aligned} \tag{DP_3}$$

Let the reduced problem

$$\begin{aligned} &p(t,u)u'^2 + g(t,u) = 0, \quad a < t < t_L, \\ &u(a) = A \end{aligned} \tag{R_L}$$

and the reduced problem

$$\begin{aligned} &p(t,u)u'^2 + g(t,u) = 0, \quad t_R < t < b, \\ &u(b) = B \end{aligned} \tag{R_L}$$

have, respectively, solutions $u = u_L(t)$ and $u = u_R(t)$ such that $t_L > t_R$, $u_L(t_0) = u_R(t_0)$, and $u_L'(t_0) \neq u_R'(t_0)$ at a point t_0 in (t_R, t_L).

We define the reduced path

$$u_0(t) = \begin{cases} u_L(t), & a \leq t \leq t_0, \\ u_R(t), & t_0 \leq t \leq b, \end{cases}$$

and we state a result due to Haber and Levinson [27] who studied a more general problem.

Theorem 5.15 (Haber and Levinson). Assume that the reduced path $u = u_0(t)$ is strongly stable in $[a,b]$ and that

$$p(t_0,\sigma)w^2 + g(t_0,\sigma) \begin{cases} > 0, \text{ if } u_L'(t_0) < w < u_R'(t_0) \\ < 0, \text{ if } u_R'(t_0) < w < u_L'(t_0), \end{cases} \qquad (*)$$

where $\sigma = u_L(t_0) = u_R(t_0)$. Then there exists an $\varepsilon_0 > 0$ such that for $0 < \varepsilon \leq \varepsilon_0$, the problem (DP_3) has a solution $y = y(t,\varepsilon)$ in $[a,b]$ satisfying

$$y(t,\varepsilon) = u_0(t) + O(\varepsilon).$$

The following corollary holds if $|p(t,y)| \geq \nu > 0$ in $\mathcal{R}(u_L,u_R)$, since the inequality (*) is automatically satisfied.

Corollary 5.15. Assume that the reduced path $u = u_0(t)$ is strongly stable in $[a,b]$ and that there is a constant $\nu > 0$ such that $|p(t,y)| \geq \nu$ in $\mathcal{R}(u_L,u_R)$. Then the conclusion of Theorem 5.15 is valid.

If the reduced path u_0 is only weakly stable in $[a,b]$, it is still possible to prove results analogous to Theorem 5.15 provided that u_0 is also (I_q)- or (II_n)-stable. We have the following two theorems, which can be proved in the same manner as Theorems 3.9 - 3.10.

Theorem 5.16. Assume that the reduced path $u = u_0(t)$ is weakly stable and (I_q)-stable in $[a,b]$. Then there exists an $\varepsilon_0 > 0$ such that for $0 < \varepsilon \leq \varepsilon_0$ the problem (DP_3) has a solution $y = y(t,\varepsilon)$ in $[a,b]$ satisfying

$$y(t,\varepsilon) = u_0(t) + O(v_I(t,\varepsilon)) + O(\varepsilon^{1/(2q+1)}),$$

where v_I is as defined in Theorem 3.9 with $|u'(t_0^+)-u'(t_0^-)|$ replaced by $|u_R'(t_0) - u_L'(t_0)|$.

Theorem 5.17. Assume that the reduced path $u = u_0(t)$ is weakly stable and (II_n)-stable in $[a,b]$. Assume also that $u_L'' \geq 0$ in (a,t_L), $u_R'' \geq 0$ in (t_R,b) and $u_L'(t_0) < u_R'(t_0)$. Then there exists an $\varepsilon_0 > 0$ such that for $0 < \varepsilon \leq \varepsilon_0$ the problem (DP_3) has a solution $y = y(t,\varepsilon)$ in $[a,b]$ satisfying

$$0 \leq y(t,\varepsilon) - u_0(t) \leq v_I(t,\varepsilon) + c\varepsilon^{1/n},$$

where v_I is as defined in Theorem 3.10 with $|u'(t_0^+) - u'(t_0^-)|$ replaced by $|u_R'(t_0) - u_L'(t_0)|$.

If the function u_0 is (III_n)-stable in $[a,b]$, then an analogous result holds, provided that $u_L'' \leq 0$ in (a,t_L), $u_R'' \leq 0$ in (t_R,b) and $u_L'(t_0) > u_R'(t_0)$.

We suppose next that the functions u_L and u_R intersect a third solution u of the reduced equation (R) at points t_1 in (a,t_L) and t_2 in (t_R,b), respectively, with $t_L < t_R$. The next two theorems relate to the reduced path

$$u = u_1(t) = \begin{cases} u_L(t), & a \leq t \leq t_1, \\ u(t), & t_1 \leq t \leq t_2, \\ u_R(t), & t_2 \leq t \leq b. \end{cases}$$

Theorem 5.18. Assume that the reduced problems (R_L), (R) and (R_R) have solutions $u = u_L(t)$, $u = u(t)$ and $u = u_R(t)$ of class $C^{(2)}([a,t_1])$, $C^{(2)}([t_1,t_2])$ and $C^{(2)}([t_2,b])$, respectively, such that $u_L(t_1) = u(t_1)$, $u_L'(t_1) \neq u'(t_1)$, $u(t_2) = u_R(t_2)$ and $u'(t_2) \neq u_R'(t_2)$ at points $t_1 < t_2$ in (a,b). Assume also that the reduced path $u = u_1(t)$ is strongly stable and (I_q)-stable in $[a,b]$. Then there exists an $\varepsilon_0 > 0$ such that for $0 < \varepsilon \leq \varepsilon_0$ the problem (DP_3) has a solution $y = y(t,\varepsilon)$ in $[a,b]$ satisfying

$$|y(t,\varepsilon) - u_1(t)| \leq v_1(t,\varepsilon) + v_2(t,\varepsilon) + c\varepsilon^{1/(2q+1)},$$

where v_1 and v_2 are as defined in Theorem 4.21.

Theorem 5.19. Assume that the reduced path $u = u_1(t)$ is strongly stable and (II_n)-stable in $[a,b]$. Assume also that $u_L'' \geq 0$ in (a,t_1), $u'' \geq 0$ in (t_1,t_2), $u_R'' \geq 0$ in (t_2,b) and $u_L'(t_1) < u'(t_1)$, $u'(t_2) < u_R'(t_2)$. Then there exists an $\varepsilon_0 > 0$ such that for $0 < \varepsilon \leq \varepsilon_0$ the problem (DP_3) has a solution $y = y(t,\varepsilon)$ in $[a,b]$ satisfying

$$0 \leq y(t,\varepsilon) - u_1(t) \leq v_1(t,\varepsilon) + v_2(t,\varepsilon) + c\varepsilon^{1/n},$$

where v_1 and v_2 are as defined in Theorem 4.22.

If the reduced path u_1 is strongly stable and (III_n)-stable in $[a,b]$, then the corresponding result is valid, provided that $u_L'' \leq 0$ in (a,t_1), $u'' \leq 0$ in (t_1,t_2), $u_R'' \leq 0$ in (t_2,b), $u_L'(t_1) > u'(t_1)$ and $u'(t_2) > u_R'(t_2)$.

Suppose now that the reduced path $u = u_1(t)$ is of class $C^{(1)}[a,b]$, that is, $u_L'(t_1) = u'(t_1)$ and $u'(t_2) = u_R'(t_2)$. This is the case if the function u is a singular solution of (R). Then results analogous to Theorems 5.18 and 5.19 hold, and are indeed their corollaries.

<u>Theorem 5.20</u>. Assume that the reduced path $u = u_1(t)$ is of class $C^{(1)}([a,b])$, weakly stable and (I_q)-stable in $[a,b]$. Then there exists an $\varepsilon_0 > 0$ such that the problem (DP_3) has a solution $y = y(t,\varepsilon)$ whenever $0 < \varepsilon \leq \varepsilon_0$. In addition, for t in $[a,b]$ we have that

$$|y(t,\varepsilon) - u_1(t)| \leq c\varepsilon^{1/(2q+1)}.$$

<u>Theorem 5.21</u>. Assume that the reduced path $u = u_1(t)$ is of class $C^{(1)}([a,b])$, weakly stable and (II_n)-stable in $[a,b]$. Assume also that $u_L'' \geq 0$ in (a,t_1), $u'' \geq 0$ in (t_1,t_2) and $u_R'' \geq 0$ in (t_2,b). Then there exists an $\varepsilon_0 > 0$ such that the problem (DP_3) has a solution $y = y(t,\varepsilon)$ whenever $0 < \varepsilon \leq \varepsilon_0$. In addition, for t in $[a,b]$ we have that

$$0 \leq y(t,\varepsilon) - u_1(t) \leq c\varepsilon^{1/n}.$$

We note that if the $C^{(1)}$-path u_1 is (III_n)-stable, then the result corresponding to Theorem 5.21 holds, if $u_1'' \leq 0$ in (a,b). In addition, if $u_L'(t_1) \neq u'(t_1)$ and $u'(t_2) = u_R'(t_2)$ or $u_L'(t_1) = u'(t_1)$ and $u'(t_2) \neq u_R'(t_2)$ then the obvious results are valid and may be obtained directly from Theorems 5.18 and 5.19.

Let us next give some results for the reduced path

$$u_2(t) = \begin{cases} u(t), & a \leq t \leq t_2, \\ u_R(t), & t_2 \leq t \leq b. \end{cases}$$

These results can be applied to the path

$$u_3(t) = \begin{cases} u_L(t), & a \leq t \leq t_1, \\ u(t), & t_1 \leq t \leq b, \end{cases}$$

by making the change of variable $t \to a + b - t$. Since the function u is not required to satisfy the boundary condition at $t = a$, we anticipate that solutions of (DP_3) described by u_2 will exhibit boundary layer behavior there, as well as interior layer behavior at $t = t_2$. The results which follow are a combination of our previous results on boundary layers and interior layers, and so their proofs are omitted.

Theorem 5.22. Assume that the reduced problems (R) and (R_R) have solutions $u = u(t)$ and $u = u_R(t)$ of class $C^{(2)}([a,t_2])$ and $C^{(2)}([t_2,b])$, respectively. Assume also that the path $u = u_2(t)$ is strongly stable and (I_q)- or ($\tilde{I}_q$)-stable in $[a,b]$, and that $p(t,y) \geq \nu > 0$ in $R^+(u,u_R) \cap [a,a+\delta]$, if $u(a) \leq A$, and $p(t,y) \leq -\nu < 0$ in $R^-(u,u_R) \cap [a,a+\delta]$, if $u(a) \geq A$, for a positive constant ν. Then there exists an $\varepsilon_0 > 0$ such that for $0 < \varepsilon \leq \varepsilon_0$ the problem (DP_3) has a solution $y = y(t,\varepsilon)$ in $[a,b]$ satisfying

(a) $-c\varepsilon^{1/(2q+1)} \leq y(t,\varepsilon) - u_2(t) \leq w_L(t,\varepsilon) + v_2(t,\varepsilon) + c\varepsilon^{1/(2q+1)}$

if $u(a) \leq A$ and $u'(t_2) < u_R'(t_2)$;

(b) $-v_2(t,\varepsilon) - c\varepsilon^{1/(2q+1)} \leq y(t,\varepsilon) - u_2(t) \leq w_L(t,\varepsilon) + c\varepsilon^{1/(2q+1)}$

if $u(a) \leq A$ and $u'(t_2) > u_R'(t_2)$;

(c) $-w_L(t,\varepsilon) - c\varepsilon^{1/(2q+1)} \leq y(t,\varepsilon) - u_2(t) \leq v_2(t,\varepsilon) + c\varepsilon^{1/(2q+1)}$

if $u(a) \geq A$ and $u'(t_2) < u_R'(t_2)$;

and

(d) $-w_L(t,\varepsilon) - v_2(t,\varepsilon) - c\varepsilon^{1/(2q+1)} \leq y(t,\varepsilon) - u_2(t) \leq c\varepsilon^{1/(2q+1)}$

if $u(a) \geq A$ and $u'(t_2) > u_R'(t_2)$,

where w_L is as defined in the conclusion of Theorem 5.1 and

$$v_2(t,\varepsilon) = \frac{1}{2}\,\varepsilon k^{-1}|u'(t_2) - u_R'(t_2)|\exp[-k\varepsilon^{-1}|t-t_2|].$$

Theorem 5.23. Assume that the reduced path $u = u_2(t)$ is strongly stable and (II_n)- or ($\tilde{II}_n$)-stable in $[a,b]$. Assume also that $u(a) \leq A$, $p(t,y) \geq \nu > 0$ in $R^+(u,u_R)$ for a positive constant ν, $u'(t_2) < u_R'(t_2)$, $u'' \geq 0$ in (a,t_2) and $u_R'' \geq 0$ in (t_2,b). Then there exists an $\varepsilon_0 > 0$ such that for $0 < \varepsilon \leq \varepsilon_0$ the problem (DP_3) has a solution $y = y(t,\varepsilon)$ in $[a,b]$ satisfying

$$0 \leq y(t,\varepsilon) - u_2(t) \leq w_L(t,\varepsilon) + v_2(t,\varepsilon) + c\varepsilon^{1/n},$$

where w_L and v_2 are as defined in the conclusion of Theorem 5.22.

Theorem 5.24. Assume that the reduced path $u = u_2(t)$ is strongly stable and (III_n)- or $(\tilde{III}_n)$-stable in $[a,b]$. Assume also that $u(a) \geq A$, $p(t,y) \leq -\nu < 0$ in $R^-(u,u_R) \cap [a,a+\delta]$ for a positive constant ν, $u'(t_2) > u_R'(t_2)$, $u'' \leq 0$ in (a,t_2) and $u_R'' \leq 0$ in (t_2,b). Then there exists an $\varepsilon_0 > 0$ such that for $0 < \varepsilon \leq \varepsilon_0$ the problem (DP_3) has a solution $y = y(t,\varepsilon)$ in $[a,b]$ satisfying

$$-w_L(t,\varepsilon) - v_2(t,\varepsilon) - c\varepsilon^{1/n} \leq y(t,\varepsilon) - u_2(t) \leq 0,$$

where w_L and v_2 are as defined in the conclusion of Theorem 5.22.

If the reduced path u_2 is weakly stable and of class $C^{(1)}([a,b])$, that is, if $u'(t_2) = u_R'(t_2)$, then the results corresponding to Theorems 5.22 - 5.24 are valid mutatis mutandis. We note that if $u'(t_2) = u_R'(t_2)$, then $v_2(t,\varepsilon) \equiv 0$.

Lastly we state our results on the interior crossing phenomena for solutions of the Robin problem (RP_6). The results relating to the problem (RP_5) are left to the reader to formulate.

Theorem 5.25. Assume that the reduced problems $(\tilde{R}_L)$ and $(\tilde{R}_R)$ have solutions $u = u_L(t)$ and $u = u_R(t)$ of class $C^{(2)}([a,t_L])$ and $C^{(2)}([t_R,b])$, respectively, such that $t_L > t_R$, $u_L(t_0) = u_R(t_0)$ $(= \sigma)$ and $u_L'(t_0) \neq u_R'(t_0)$ at a point t_0 in (t_R,t_L). Assume also that the reduced path

$$u = u_0(t) = \begin{cases} u_L(t), & a \leq t \leq t_0, \\ u_R(t), & t_0 \leq t \leq b, \end{cases}$$

is strongly stable and (I_q)-stable in $[a,b]$, and that

$$p(t_0,\sigma)w^2 + g(t_0,\sigma) \begin{cases} > 0, \text{ for } u_L'(t_0) < w < u_R'(t_0), \\ < 0, \text{ for } u_R'(t_0) < w < u_L'(t_0). \end{cases} \qquad (*)$$

Then there exists an $\varepsilon_0 > 0$ such that for $0 < \varepsilon \leq \varepsilon_0$ the problem (RP_6) has a solution $y = y(t,\varepsilon)$ in $[a,b]$ satisfying

$$y(t,\varepsilon) = u_0(t) + O(\varepsilon^{1/(2q+1)}).$$

We note that if $|p(t,y)| > 0$ in $R(u_L,u_R)$, then the inequality (*) is automatically satisfied.

Theorems 5.16 - 5.21 hold for the problem (RP_6) mutatis mutandis, and so we proceed directly to a consideration of the analogs of Theorems 5.22 - 5.24 for the reduced path

$$u = u_2(t) = \begin{cases} u(t), & a \leq t \leq t_2, \\ u_R(t), & t_2 \leq t \leq b. \end{cases}$$

Theorem 5.26. Assume that the reduced problems (R) and $(\tilde{R}_R)$ have solutions $u = u(t)$ and $u = u_R(t)$ of class $C^{(2)}([a,t_2])$ and class $C^{(2)}([t_2,b])$, respectively. Assume also that the path $u = u_2(t)$ is strongly stable and (I_q)-stable in $[a,b]$, and that $u(a) - p_1u'(a) = A$ or

$$(u(a) - p_1u'(a) - A)[p(a,u(a))\{p_1^{-1}(A-u(a))\}^2 + g(a,u(a))] > 0.$$

Then there exists an $\varepsilon_0 > 0$ such that for $0 < \varepsilon \leq \varepsilon_0$ the problem (RP_6) has a solution $y = y(t,\varepsilon)$ in $[a,b]$ satisfying

$$|y(t,\varepsilon) - u_2(t)| \leq v_L(t,\varepsilon) + v_2(t,\varepsilon) + c\varepsilon^{1/(2q+1)},$$

where v_L is as defined in the conclusion of Theorem 5.9 with u_R replaced by u, and v_2 is as defined in the conclusion of Theorem 5.22.

Theorem 5.27. Assume that the reduced path $u = u_2(t)$ is strongly stable and (II_n)-stable in $[a,b]$. Assume also that $u(a) - p_1u'(a) \leq A$, $p(t,y) \geq 0$ in $R^+(u,u_R) \cap [a,a+\delta]$, $u'(t_2) < u_R'(t_2)$, $u'' \geq 0$ in (a,t_2) and $u_R'' \geq 0$ in (t_2,b). Then there exists an $\varepsilon_0 > 0$ such that for $0 < \varepsilon \leq \varepsilon_0$ the problem (RP_6) has a solution $y = y(t,\varepsilon)$ in $[a,b]$ satisfying

$$0 \leq y(t,\varepsilon) - u_2(t) \leq v_L(t,\varepsilon) + v_2(t,\varepsilon) + c\varepsilon^{1/n},$$

where v_L and v_2 are as defined in the conclusion of Theorem 5.27.

Theorem 5.28. Assume that the reduced path $u = u_2(t)$ is strongly stable and (III_n)-stable in $[a,b]$. Assume also that

$$u(a) - p_1u'(a) \geq A, \quad p(t,y) \leq 0 \quad \text{in } R^-(u,u_R) \cap [a,a+\delta],$$

$u'(t_2) > u_R'(t_2)$, $u'' \leq 0$ in (a,t_2) and $u_R'' \leq 0$ in (t_2,b). Then there exists an $\varepsilon_0 > 0$ such that for $0 < \varepsilon \leq \varepsilon_0$ the problem (RP_6) has a solution $y = y(t,\varepsilon)$ in $[a,b]$ satisfying

$$-v_L(t,\varepsilon) - v_2(t,\varepsilon) - c\varepsilon^{1/n} \leq y(t,\varepsilon) - u_2(t) \leq 0,$$

where v_L and v_2 are as defined in the conclusion of Theorem 5.26.

We note that if the path u_2 is $C^{(1)}$-smooth, that is, if $u'(t_2) = u_R'(t_2)$, and weakly stable, then the results corresponding to Theorems 5.26 and 5.28 hold mutatis mutandis with $v_2 \equiv 0$.

Notes and Remarks

5.1. The theory developed above for the problems (DP_3), (RP_5) and (RP_6) can accommodate, with minor modification, the more general problem in which $p(t,y)$, $g(t,y)$, A and B are functions of ε satisfying

$$p(t,y,\varepsilon) = p(t,y,0) + o(1), \quad g(t,y,\varepsilon) = g(t,y,0) + o(1),$$

$$A(\varepsilon) = A(0) + o(1) \quad \text{and} \quad B(\varepsilon) = B(0) + o(1).$$

5.2. Surprisingly little work has been done on the Dirichlet problem (DP_3). Motivated by a brief discussion in Section 5 of the paper by Dorr, Parter and Shampine [20], Howes has developed a reasonably coherent boundary layer theory for (DP_3) in [39]. He has also discussed a corresponding interior layer theory in [39] and [38] which includes not only the interior crossing phenomena treated above but also shock layer phenomena. The classic interior crossing result of Haber and Levinson [27] has been extended in several directions by Vasil'eva [87], O'Malley [74] and Howes [43]. The Robin problems (RP_5) and (RP_6) have been considered by Vasil'eva [87], Macki [62], Searl [84] and Howes [40].

Chapter VI
Superquadratic Singular Perturbation Problems

§6.1. Introduction

In previous chapters we have presented fairly comprehensive results for boundary value problems involving the differential equation

$$\varepsilon y'' = f(t,y,y'), \quad a < t < b,$$

subject to the fundamental restriction:

$$f(t,y,z) = O(|z|^2) \quad \text{as} \quad |z| \to \infty.$$

It is natural for us to ask if similar results can be extended to these boundary value problems when f is subjected to the restriction:

$$f(t,y,z) = O(|z|^n) \quad \text{as} \quad |z| \to \infty \quad \text{for} \quad n > 2.$$

A partial and somewhat negative answer was given many years ago by Vishik and Liusternik [90] for the Dirichlet boundary value problem

$$\varepsilon y'' = f(t,y,y'), \quad a < t < b, \tag{6.1}$$

$$y(a,\varepsilon) = A, \quad y(b,\varepsilon) = B. \tag{6.2}$$

They showed that if

$$f = p(t,y)y'^n + f_1(t,y,y'),$$

where $n > 2$, $|p(t,y)| \geq \nu > 0$ and

$$f_1(t,y,z) = O(|z|^{n-1}) \quad \text{as} \quad |z| \to \infty,$$

then every solution of (6.1), (6.2), if it exists (cf. (E_{14}) below), satisfies

$$y'(a,\varepsilon) = O(1) \quad \text{and} \quad y'(b,\varepsilon) = O(1) \quad \text{as} \quad \varepsilon \to 0^+.$$

[Note that these authors actually considered the Cauchy problem (6.1), for $0 < n < 2$, together with the initial conditions $y(a,\varepsilon) = A$, $y'(a,\varepsilon) = C/\varepsilon^\gamma$, $C,\gamma > 0$. This Cauchy problem is equivalent to the above Dirichlet problem, as the reader can see by solving the simple example

$$\varepsilon y'' = -\nu y', \quad \nu > 0, \quad y(0,\varepsilon) = A, \quad y(1,\varepsilon) = B, \text{ where } A < B.$$

A short calculation shows that here $\gamma = 1$ and $C \sim \nu(B - A)$. However, when $n = 2$, they showed that this equivalence is obtained only if $y'(a,\varepsilon) \sim \varepsilon e^{k/\varepsilon}$, $k > 0$. Also compare this with our discussion at the beginning of Chapter I.]

Thus, in contrast to the solutions for similar problems studied in previous chapters, Vishik and Liusternik's problem does not exhibit a boundary layer characteristic! Another way of formulating their result is the following. Let $u = u(t)$ be a solution of the reduced equation $f(t,u,u') = 0$ which does not satisfy either boundary condition, that is, $u(a) \neq A$ and $u(b) \neq B$, then there is *no* solution $y = y(t,\varepsilon)$ of (6.1), (6.2) satisfying

$$\lim_{\varepsilon\to 0^+} y(t,\varepsilon) = u(t) \quad \text{for} \quad a < t < b.$$

This implies that for arbitrary choices of A and B, the Dirichlet problem (6.1), (6.2) has *no* solution for all sufficiently small values of ε. Therefore, the Dirichlet problem (6.1), (6.2), where $n > 2$ and $f = O(|y'|^n)$ as $|y'| \to \infty$ is not, in general, a well-posed problem, since we are interested in existence of solutions for small values of ε. These remarks are best illustrated by the classic counterexample due to Coddington and Levinson [14] (see also [23], [94]):

$$\begin{aligned} &\varepsilon y'' = -y' - y'^3 \equiv f(y'), \quad 0 < t < 1, \\ &y(0,\varepsilon) = A, \quad y(1,\varepsilon) = B. \end{aligned} \qquad (E_{14})$$

By quadratures we obtain the general solution

$$y(t,\varepsilon) = \pm\varepsilon \arcsin (e^{(x+c_1)/\varepsilon}) + c_2,$$

where c_1, c_2 are arbitrary constants. To choose these constants to satisfy the boundary conditions, we run into a difficulty. For all sufficiently small values of ε, in fact, if $0 < \varepsilon < 2|A - B|/\pi$, we have

$$|y(t,\varepsilon) - A| = O(\varepsilon),$$

and so $y(1,\varepsilon)$ cannot be equal to B, unless $B = A$. Thus there is no solution, unless $B = A$. In this latter case, we obtain the constant solution $y(t,\varepsilon) \equiv A = B$.

For this counterexample, note that

$$f(y') = O(|y'|^3), \text{ as } |y'| \to \infty,$$

and that the only real solution of the reduced equation $f(u') = 0$ is $u(t) \equiv$ constant. This reduced solution $u(t) \equiv$ constant cannot satisfy both boundary conditions, unless $A = B$, and so there is no solution to (E_{14}) as $\varepsilon \to 0^+$, unless $A = B$, which is in agreement with the result of Vishik and Liusternik.

The above results seem to indicate that there is very little that one can do with Dirichlet boundary value problems. However, we hasten to point out that, fortunately, the Robin problem for this class of differential equations turns out to be well-posed and solvable. That this is so should not be surprising, in view of our discussion of it in Chapter V, and also in view of the result of Vishik and Liusternik, which implies that the solution $y(t,\varepsilon)$ for the *Robin* boundary value problem should satisfy

$$y'(a,\varepsilon) = O(1) = p_1^{-1}(y(a,\varepsilon) - A)$$

and

$$y'(b,\varepsilon) = O(1) = p_2^{-1}(y(b,\varepsilon) - B).$$

This will be the case if $y(a,\varepsilon)$ and $y(b,\varepsilon)$ are $O(1)$ as $\varepsilon \to 0^+$. Thus, solutions of such problems have uniformly bounded derivatives at the endpoints.

§6.2. A Dirichlet Problem

To study these *superquadratic* boundary value problems in detail, we first consider the following class of Dirichlet problems

$$\begin{aligned} &\varepsilon y'' = h(t,y)f(t,y,y') \equiv F(t,y,y'), \quad a < t < b \\ &y(a,\varepsilon) = A, \quad y(b,\varepsilon) = B. \end{aligned} \qquad (DP_4)$$

Here the function $h(t,y)$ is of the same type as in Chapter III, while the function f is continuous and satisfies $f(t,y,z) \geq \mu > 0$ for (t,y) in a domain $\mathcal{D}_0(u)$ and for all z in $\mathbb{R}$. The domain $\mathcal{D}_0(u)$ is as defined in Chapter III, where $u = u(t)$ is a solution of the reduced equation

$$h(t,u) = 0, \quad a \leq t \leq b. \tag{R_4}$$

The stability properties of $u(t)$ are as given in Definitions 3.1, 3.3 and 3.3, with m replaced by $m\mu^{-1}$. For example, if $u(t)$ is (I_0)-stable, then $h_y(t,y) \geq m\mu^{-1} > 0$ for (t,y) in $\mathcal{D}_0(u)$. We also assume that

$$f(t,y,z) = O(|z|^n) \quad \text{as} \quad |z| \to \infty \quad \text{for} \quad n > 2.$$

We note that for $n = 2$, the results of Chapter III for the Dirichlet problem apply directly to the problem (DP_4), provided that the stability constant m is replaced by $m\mu^{-1}$.

The following two theorems give the basic results for the Dirichlet problem (DP_4).

Theorem 6.1. Assume that the reduced equation (R_4) has an (I_q)- (or (II_n)-) stable solution $u = u(t)$ of class $C^{(2)}([a,b])$ such that $u(a) = A$, $u(b) = B$ and $u'' \geq 0$ in (a,b). Then there exists an $\varepsilon_0 \geq 0$ such that for $0 < \varepsilon \leq \varepsilon_0$ the problem (DP_4) has a solution $y = y(t,\varepsilon)$ in $[a,b]$ satisfying

$$u(t) \leq y(t,\varepsilon) \leq u(t) + c\varepsilon^{1/\rho},$$

where $\rho = 2q + 1$ (or n), and c is a known positive constant depending on m, $|u''|$ and ρ.

Proof: Define for t in $[a,b]$ and $\varepsilon > 0$

$$\alpha(t,\varepsilon) = u(t), \quad \beta(t,\varepsilon) = u(t) + \Gamma(\varepsilon),$$

where $\Gamma(\varepsilon) = (\varepsilon\gamma m^{-1})^{1/\rho}$, for $\gamma \geq \rho!\, M$ and $M \geq |u''|$. Clearly, $\alpha \leq \beta$, $\alpha(a,\varepsilon) \leq A \leq \beta(a,\varepsilon)$, $\alpha(b,\varepsilon) \leq B \leq \beta(b,\varepsilon)$, and $\varepsilon\alpha'' \geq F(t,\alpha,\alpha')$ (since $u'' \geq 0$). It only remains to show that $\varepsilon\beta'' \leq F(t,\beta,\beta')$. We have

$$\begin{aligned} F(t,\beta,\beta') - \varepsilon\beta'' &= h(t,\beta)f(t,\beta,\beta') - \varepsilon\beta'' \\ &= \left[h(t,u) + \sum_{j=1}^{\rho-1} \partial_y^i h(t,u)\Gamma^j/j! + \partial_y^\rho h(t,\xi)\Gamma^\rho/\rho!\right] f(t,\beta,\beta') - \varepsilon u'' \\ &\geq (m\mu^{-1}\varepsilon\gamma m^{-1}/\rho!)\mu - \varepsilon M \geq 0, \end{aligned}$$

since $\gamma \geq \rho!\, M$. Here $\xi = \xi(t,\varepsilon) = u(t) + \theta(\beta - u(t))$, $0 < \theta < 1$, and (t,ξ) belongs to $\mathcal{D}_0(u)$ for all sufficiently small ε, say $0 < \varepsilon \leq \varepsilon_0$.

Consequently, α and β satisfy all of the required inequalities. The proof will follow from Theorem 2.1 if we can show that whenever

$$\varepsilon y'' = F(t,y,y') \quad \text{and} \quad \alpha \le y \le \beta \quad \text{on} \quad J \subset [a,b],$$

then $|y'(t,\varepsilon)| \le N$ on $J \times (0,\varepsilon_0]$. Indeed, the solution $y(t,\varepsilon)$ satisfies $u'(a) \le y'(t,\varepsilon) \le u'(b)$. This is because $y(a,\varepsilon) = u(a) = \alpha(a,\varepsilon)$ and $\alpha(t,\varepsilon) \le y(t,\varepsilon)$ imply that $y'(a,\varepsilon) \ge \alpha'(a,\varepsilon) = u'(a)$, while $y(b,\varepsilon) = u(b) = \alpha(b,\varepsilon)$ and $\alpha(t,\varepsilon) \le y(t,\varepsilon)$ imply that $y'(b,\varepsilon) \le \alpha'(b,\varepsilon) = u'(b)$. Therefore $u'(a) \le y'(t,\varepsilon) \le u'(b)$ since $y'' \ge 0$ for $\alpha \le y \le \beta$.

Theorem 6.2 is the "*concave*" version of Theorem 6.1.

Theorem 6.2. Assume that the reduced equation (R_4) has an (I_q)- or (III_n)-stable solution $u = u(t)$ of class $C^{(2)}([a,b])$ such that $u(a) = A$, $u(b) = B$ and $u'' \le 0$ in (a,b). Then there exists an $\varepsilon_0 > 0$ such that for $0 < \varepsilon \le \varepsilon_0$ the problem (DP_4) has a solution $y = y(t,\varepsilon)$ in $[a,b]$ satisfying

$$u(t) - c\varepsilon^{1/\rho} \le y(t,\varepsilon) \le u(t),$$

where the constants c and ρ are as defined in Theorem 6.1.

Proof: Simply let $y \to -y$ and apply Theorem 6.1 to the transformed problem.

We remark that since it is assumed that $u(a) = A$ and $u(b) = B$, the domain $\mathcal{D}_0(u)$ in which the functions h and f are defined is of the form $\mathcal{D}_0(u) = \{(t,y)\colon a \le t \le b,\ |y - u(t)| \le \delta\}$, where $\delta > 0$ is an arbitrarily small (but fixed) positive constant. Consequently, for such problems (I_q)-, (II_n)- and (III_n)-stability are essentially equivalent to their "tiled" counterparts.

§6.3. Robin Problems: Boundary Layer Phenomena

We turn now to the following classes of Robin problems

$$\varepsilon y'' = h(t,y)f(t,y,y'), \quad a < t < b,$$
$$y(a,\varepsilon) - p_1 y'(a,\varepsilon) = A, \quad y(b,\varepsilon) = B, \qquad (RP_7)$$

and

$$\varepsilon y'' = h(t,y)f(t,y,y'), \quad a < t < b,$$
$$y(a,\varepsilon) - p_1 y'(a,\varepsilon) = A, \quad y(b,\varepsilon) + p_2 y'(b,\varepsilon) = B, \qquad (RP_8)$$

where p_1 and p_2 are positive constants. The analogs of Theorems 6.1 and 6.2 for these problems follow easily from our discussion of the problems (RP_1) and (RP_2) in Chapter III, and so we omit the proofs of the next two theorems.

Theorem 6.3. Assume that the reduced equation (R_4) has an (I_q)- or (II_n)-stable solution $u = u(t)$ of class $C^{(2)}([a,b])$ such that $u(a) - p_1u'(a) \leq A$, $u(b) = B$ and $u'' \geq 0$ in (a,b). Then there exists an $\varepsilon_0 > 0$ such that for $0 < \varepsilon \leq \varepsilon_0$ the problem (RP_7) has a solution $y = y(t,\varepsilon)$ in $[a,b]$ satisfying

$$u(t) \leq y(t,\varepsilon) \leq u(t) + v_L(t,\varepsilon) + c\varepsilon^{1/\rho}.$$

Here c is as defined in Theorem 6.1, and $v_L(t,\varepsilon)$ is as defined in Theorems 3.4 and 3.5 for $\rho = 2q+1$ and $\rho = n$, respectively.

The following result refers to the analogous problem (RP_8).

Theorem 6.4. Assume that the reduced equation (R_4) has an (I_q)- or (II_n)-stable solution $u = u(t)$ of class $C^{(2)}([a,b])$ such that $u(a) - p_1u'(a) \leq A$, $u(b) + p_2u'(b) \leq B$ and $u'' \geq 0$ in (a,b). Then there exists an $\varepsilon_0 > 0$ such that for $0 < \varepsilon \leq \varepsilon_0$ the problem (RP_8) has a solution $y = y(t,\varepsilon)$ in $[a,b]$ satisfying

$$u(t) \leq y(t,\varepsilon) \leq u(t) + v_L(t,\varepsilon) + v_R(t,\varepsilon) + c\varepsilon^{1/\rho},$$

where c and v_L are as in Theorem 6.3 and v_R is as defined in Theorems 3.7 and 3.8 for $\rho = 2q + 1$ and $\rho = n$, respectively.

If the function u is (I_q)- or (III_n)-stable and satisfies the reverse inequalities, then results analogous to Theorems 6.3 and 6.4 hold true. Tiis can be seen by using the change of variable $y \to -y$ and applying Theorems 6.3 and 6.4, respectively.

§6.4. Interior Layer Phenomena

It is now an easy matter to consider the occurrence of interior layer behavior for the problems (DP_4), (RP_7) and (RP_8). The situation described in the following theorems arises most frequently when two solutions of the reduced equation (R_4) intersect at a point t_0 in (a,b) with unequal slopes, as we have discussed earlier in Chapter III. We omit the straightforward proofs.

Theorem 6.5. Assume that the reduced equation (R_4) has an (I_q)- or (II_n)-stable solution $u = u(t)$ of class $C^{(2)}([a,b])$, except at the point t_0

in (a,b) where $u'(t_0^-) < u'(t_0^+)$ and $|u''(t_0^\pm)| < \infty$. Assume also that $u(a) = A$, $u(b) = B$ and $u'' \geq 0$ in $(a,t_0) \cup (t_0,b)$. Then there exists an $\varepsilon_0 > 0$ such that for $0 < \varepsilon \leq \varepsilon_0$ the problem (DP_4) has a solution $y = y(t,\varepsilon)$ in $[a,b]$ satisfying

$$u(t) \leq y(t,\varepsilon) \leq u(t) + v_I(t,\varepsilon) + c\varepsilon^{1/\rho},$$

where c is a positive constant and v_I is defined in Theorems 3.9 and 3.10 for $\rho = 2q+1$ and $\rho = n$, respectively.

It is clear that a result analogous to Theorem 6.5 can be obtained if the solution u is (I_q)- (or (III_n)-) stable, and

$$u'(t_0^-) > u'(t_0^+) \quad \text{and} \quad u'' \leq 0 \text{ in } (a,t_0) \cup (t_0,b).$$

We leave its exact formulation to the reader.

In the same manner we can prove the next two results which deal with 'interior crossing' phenomena for the Robin problems (RP_7) and (RP_8), when the reduced solution u is (I_q)- (or (II_n)-) stable and *convex*. Similar results can be obtained when u is (I_q)- (or (III_n)-) stable and *concave*.

Theorem 6.6. Assume that the reduced equation (R_4) has an (I_q)- or (II_n)-stable solution $u = u(t)$ of class $C^{(2)}([a,b])$, except at the point t_0 in (a,b) where $u'(t_0^-) < u'(t_0^+)$ and $|u''(t_0^\pm)| < \infty$. Assume also that $u(a) - p_1u'(a) \leq A$, $u(b) = B$ and $u'' \geq 0$ in $(a,t_0) \cup (t_0,b)$. Then there exists an $\varepsilon_0 > 0$ such that for $0 < \varepsilon \leq \varepsilon_0$ the problem (RP_7) has a solution $y = y(t,\varepsilon)$ in $[a,b]$ satisfying

$$u(t) \leq y(t,\varepsilon) \leq u(t) + v_L(t,\varepsilon) + v_I(t,\varepsilon) + c\varepsilon^{1/\rho},$$

where c and v_L are as given in Theorem 6.3 and v_I is as given in Theorem 6.5.

Theorem 6.7. Assume that the reduced equation (R_4) has an (I_q)- or (II_n)-stable solution $u = u(t)$ of class $C^{(2)}([a,b])$, except at the point t_0 in (a,b) where $u'(t_0^-) < u'(t_0^+)$ and $|u''(t_0^\pm)| < \infty$. Assume also that $u(a) - p_1u'(a) \leq A$, $u(b) + p_2u'(b) \leq B$ and $u'' \geq 0$ in $(a,t_0) \cup (t_0,b)$. Then there exists an $\varepsilon_0 > 0$ such that for $0 < \varepsilon \leq \varepsilon_0$ the problem (RP_8) has a solution $y = y(t,\varepsilon)$ in $[a,b]$ satisfying

$$u(t) \leq y(t,\varepsilon) \leq u(t) + v_L(t,\varepsilon) + v_I(t,\varepsilon) + v_R(t,\varepsilon) + c\varepsilon^{1/\rho},$$

where c and v_I are as given in Theorem 6.6 and v_R is as in Theorem 6.4.

§6.5. A General Dirichlet Problem

We now return to the general equation given at the beginning of this chapter, namely,

$$\varepsilon y'' = f(t,y,y'), \quad a < t < b.$$

In the sequel it is assumed that

$$f(t,y,z) = O(|z|^n) \quad \text{as} \quad |z| \to \infty \quad \text{for} \quad n > 2.$$

Let us first give two definitions of stability for the solution of the reduced equation

$$f(t,u,u') = 0. \tag{R}$$

They are obvious extensions of earlier ones.

Definition 6.1. A solution $u = u(t)$ of the reduced equation (R) is said to be (I_q)-, (II_n)- or (III_n)-stable in $[a,b]$ if the respective inequalities in Definitions 3.1-3.3 hold for the function $h(t,y) = f(t,y,u'(t))$ in $\mathcal{D}_0(u) = \{(t,y): a \leq t \leq b, |y - u(t)| \leq \delta\}$.

Definition 6.2. A solution $u = u(t)$ of (R) is said to be stable in $[a,b]$ if there is a positive constant k such that $|f_{y'}(t,y,y')| \geq k > 0$ in $\mathcal{D}_1(u) = \{(t,y,y'): a \leq t \leq b, |y - u(t)| \leq \delta, |y' - u'(t)| \leq \delta\}$.

With these stability properties we can now discuss the Dirichlet problem

$$\begin{aligned} &\varepsilon y'' = f(t,y,y'), \quad a < t < b, \\ &y(a,\varepsilon) = A, \quad y(b,\varepsilon) = B, \end{aligned} \tag{DP_5}$$

and we obtain the same results as Theorems 6.1 and 6.2.

Theorem 6.8. Assume that the reduced equation (R) has an (I_q)- or (II_n)-stable solution $u = u(t)$ of class $C^{(2)}([a,b])$ such that $u(a) = A$, $u(b) = B$ and $u'' \geq 0$ in $[a,b]$. Then there exists an $\varepsilon_0 > 0$ such that for $0 < \varepsilon \leq \varepsilon_0$ the problem (DP_5) has a solution $y = y(t,\varepsilon)$ in $[a,b]$ satisfying

$$u(t) \leq y(t,\varepsilon) \leq u(t) + c\varepsilon^{1/\rho},$$

where $\rho = 2q + 1$ (or n) and c is a known positive constant depending on m, $|u''|$ and ρ.

Proof: We proceed as in the proof of Theorem 6.1 by defining the same bounding pair

$$\alpha(t,\varepsilon) = u(t), \quad \beta(t,\varepsilon) = u(t) + \Gamma(\varepsilon),$$

where $\Gamma(\varepsilon) = (\varepsilon\gamma\, m^{-1})^{1/\rho}$ for $\gamma \geq \rho!\ |u''|$. Then it is clear that $\alpha \leq \beta$, $\alpha(a,\varepsilon) = A \leq \beta(a,\varepsilon)$, $\alpha(b,\varepsilon) = B \leq \beta(b,\varepsilon)$, $\varepsilon\alpha'' = \varepsilon u'' \geq f(t,\alpha,\alpha') = 0$, and $\varepsilon\beta'' \leq f(t,\beta,\beta')$. This last inequality follows by virtue of our stability assumption and our choice of γ, that is,

$$\begin{aligned} f(t,\beta,\beta') - \varepsilon\beta'' &= f(t,u,u') + \sum_{j=1}^{\rho-1} \partial_j^i f(t,u,u')\beta^j/j! \\ &\quad + \partial_y^\rho f(t,\xi,u')\beta^\rho/\rho! - \varepsilon u'' \\ &\geq m(\varepsilon\gamma m^{-1})/\rho! - \varepsilon|u''| \\ &\geq 0. \end{aligned}$$

Finally we can show by arguing as in the proof of Theorem 6.1 that the function f satisfies a generalized Nagumo condition with respect to α and β, and so the conclusion of Theorem 6.8 follows from Theorem 2.1.

The next result is the *concave* version of Theorem 6.8.

Theorem 6.9. Assume that the reduced equation (R) has an (I_q)- or (III_n)-stable solution $u = u(t)$ of class $C^{(2)}([a,b])$ such that $u(a) = A$, $u(b) = B$ and $u'' \leq 0$ in $[a,b]$. Then there exists an $\varepsilon_0 > 0$ such that for $0 < \varepsilon \leq \varepsilon_0$ the problem (DP_5) has a solution $y = y(t,\varepsilon)$ in $[a,b]$ satisfying

$$u(t) - c\varepsilon^{1/\rho} \leq y(t,\varepsilon) \leq u(t),$$

where the constants c and ρ are as defined in Theorem 6.8.

A stronger result than the above two results can be obtained, if the reduced solution $u(t)$ is stable in the sense of Definition 6.2.

Theorem 6.10. Assume that the reduced equation (R) has a stable solution $u = u(t)$ of class $C^{(2)}([a,b])$ satisfying $u(a) = A$ and $u(b) = B$. Then there exists an $\varepsilon_0 > 0$ such that for $0 < \varepsilon \leq \varepsilon_0$ the problem (DP_5) has a solution $y = y(t,\varepsilon)$ in $[a,b]$ satisfying

$$|y(t,\varepsilon) - u(t)| \leq c\varepsilon,$$

where c is a known positive constant depending on f and u.

Proof: We suppose first that $f_{y'} \leq -k < 0$. Clearly, there exists an $\ell > 0$ such that $|f_y(t,y,y')| \leq \ell$ in $\mathcal{D}_1(u)$. Setting $z = y - u(t)$ as in the proof of Theorem 4.1, we are first led to the problem

$$\varepsilon z'' = f(t,u+z,\ u' + z') - \varepsilon u''$$

$$= f_y[\cdot]z + f_{y'}[\cdot\cdot]z' - \varepsilon u'',$$

$$z(a,\varepsilon) = 0, \qquad z(b,\varepsilon) = 0,$$

where $[\cdot]$ and $[\cdot\cdot]$ denote the appropriate *intermediate* points, and then to the problem

$$\varepsilon z'' + kz' + \ell z = -\varepsilon u''$$

$$z(a,\varepsilon) = 0, \quad z(b,\varepsilon) = 0.$$

If $0 < \varepsilon < k^2/4\ell$, one of the two negative roots of the corresponding auxiliary equation $\varepsilon\lambda^2 + k\lambda + \ell = 0$ is $\lambda = -\ell/k + O(\varepsilon)$. Then the function

$$\Gamma(t,\varepsilon) = \varepsilon\gamma\ell^{-1}(\exp[-\lambda(b-t)] - 1)$$

is a solution of

$$\varepsilon\Gamma'' + k\Gamma' + \ell\Gamma = -\varepsilon\gamma.$$

It has the following properties:

$$\Gamma(b) = 0, \quad 0 \le \Gamma \le c\varepsilon \quad \text{and} \quad -c\varepsilon < \Gamma' < 0,$$

for some $c > 0$.

We define the bounding pair

$$\alpha(t,\varepsilon) = u(t) - \Gamma(t,\varepsilon),$$

$$\beta(t,\varepsilon) = u(t) + \Gamma(t,\varepsilon),$$

and we need only verify that $\varepsilon\alpha'' \ge f(t,\alpha,\alpha')$, since the differential inequality for β follows by symmetry and the other required inequalities clearly hold true. By Taylor's Theorem, however, we obtain

$$\begin{aligned}\varepsilon\alpha'' - f(t,\alpha,\alpha') &= \varepsilon u'' - \varepsilon\Gamma'' + f_{y'}[\cdot\cdot]\Gamma' - f_y[\cdot]\Gamma \\ &\ge -\varepsilon|u''| - \varepsilon\Gamma'' - k\Gamma' - \ell\Gamma \\ &\ge \varepsilon(\gamma - |u''|) \ge 0\end{aligned}$$

by choosing $\gamma \ge |u''|$.

Finally, it is not difficult to see that $y'(t,\varepsilon) - u'(t) = O(\varepsilon)$ for any solution of $\varepsilon y'' = f(t,y,y')$ which satisfies $\alpha \le y \le \beta$, and so the theorem follows from Theorem 2.1. We remark that in the proof, if $f_{y'} \ge k > 0$, then we would use the bounding functions

$$\alpha(t,\varepsilon) = u(t) - \tilde{\Gamma}(t,\varepsilon),$$

$$\beta(t,\varepsilon) = u(t) + \tilde{\Gamma}(t,\varepsilon),$$

where

$$\tilde{\Gamma}(t,\varepsilon) = \varepsilon\gamma\ell^{-1}(\exp[\lambda(a - t)] - 1).$$

§6.6. A General Robin Problem: Boundary and Interior Layer Phenomena

We now turn our attention to the nonlinear Robin problem

$$\begin{aligned} &\varepsilon y'' = f(t,y,y'), \quad a < t < b, \\ &y(a,\varepsilon) - p_1 y'(a,\varepsilon) = A, \quad y(b,\varepsilon) + p_2 y'(b,\varepsilon) = B. \end{aligned} \tag{RP_9}$$

It turns out that if $p_1 > 0$ and $p_2 > 0$, and if appropriate stability assumptions hold, then these problems will have solutions, irrespective of the growth of f with respect to y'. To formulate these stability assumptions, let us suppose that the reduced problem

$$\begin{aligned} &f(t,u,u') = 0, \quad a < t < b, \\ &u(b) + p_2 u'(b) = B, \end{aligned} \tag{R_R}$$

has a smooth solution $u = u_R(t)$ which we will use to approximate the solution of (RP_9). Since, in general, $u_R(a) - p_1 u_R'(a) \neq A$, we require u_R to have two stability properties. The first requirement is, of course, that u_R is stable in the sense of Definition 6.2, that is,

$$f_{y'}(t,y,y') \leq -k < 0 \tag{6.3}$$

in the region

$$\mathcal{D}_1(u_R) = \{(t,y,y'): a \leq t \leq b, \ |y - u_R(t)| \leq \delta, \ |y' - u_R'(t)| \leq \delta\}.$$

The second requirement is new and its motivation can be seen from the stability results for the following class of initial value problems

$$\begin{aligned} &\varepsilon y'' = f(t,y,y'), \quad a < t < b, \\ &y(a,\varepsilon) = u_R(a), \quad y'(a,\varepsilon) = p_1^{-1}(u_R(a) - A) \equiv \xi. \end{aligned} \tag{IVP}$$

Note that here $y(a,\varepsilon) - p_1 y'(a,\varepsilon) = A$. It is known (cf. [61], [87]) that the solution of (IVP) is uniformly close to the reduced solution u_R for all ξ such that either $u_R'(a) = \xi$ or

$$(u_R'(a) - \xi) f(a, u_R(a), \lambda) > 0,$$

for all λ in $(u_R'(a), \xi]$ or $[\xi, u_R'(a))$. Therefore, the second requirement is that if $u_R(a) - p_1u_R'(a) \neq A$, then

$$(u_R(a) - p_1u_R'(a) - A)f(a,u_R(a),\lambda) < 0, \tag{6.4}$$

for all λ in $(u_R'(a), p_1^{-1}(u_R(a) - A)]$ or in $[p_1^{-1}(u_R(a) - A), u_R'(a))$. This inequality provides us with the required "boundary layer stability" of the function u_R, and for nonlinear functions f, it serves to define the admissible boundary layer jump $|A - u_R(a) + p_1u_R'(a)|$.

Finally, we must ensure that u_R approximates the solution of (RP_9) at $t = b$, and so we assume that

$$p_2f_y(b,u_R(b),u_R'(b)) - f_{y'}(b,u_R(b),u_R'(b)) \neq 0. \tag{6.5}$$

Theorem 6.11. Assume that the reduced problem (R_R) has a solution $u = u_R(t)$ of class $C^{(2)}([a,b])$ satisfying the relations (6.3) - (6.5). Then there exists an $\varepsilon_0 > 0$ such that for $0 < \varepsilon \leq \varepsilon_0$ the problem (RP_9) has a solution $y = y(t,\varepsilon)$ in $[a,b]$ satisfying

$$y(t,\varepsilon) = u_R(t) + O(\varepsilon\ell(p_1k)^{-1}\exp[-k(t-a)/\varepsilon]) + O(\varepsilon),$$

where $\ell = |A - u_R(a) + p_1u_R'(a)|$.

Proof: The bounding functions are defined in the usual manner and using the standard techniques, it is not difficult to verify that the required inequalities are indeed satisfied. In order to apply Theorem 2.3, we must also verify that if $y = y(t,\varepsilon)$ is a solution of $\varepsilon y'' = f(t,y,y')$ which lies between the lower and upper solutions, then $y'(t,\varepsilon)$ is uniformly bounded. This is true, however, since $y'(t,\varepsilon) = u_R'(t) + O(\ell p_1^{-1}\exp[-k(t-a)/\varepsilon]) + O(\varepsilon)$; see [40] for complete details.

We remark that if, instead, the reduced equation $f = 0$ has a smooth solution $u = u_L(t)$ which satisfies $u_L(a) - p_1u_L'(a) = A$, then a result analogous to Theorem 6.11 can be obtained *mutatis mutandis*.

Lastly we formulate a result for the Robin problem (RP_9) which displays angular interior layer behavior which is similar to that discussed by Haber and Levinson [27] for the Dirichlet problem (cf. Chapter V). However, we must proceed with care as the following example shows.

The problem is

$$\begin{aligned} &\varepsilon y'' = 1 - y'^4, \quad 0 < t < 1, \\ &y(0,\varepsilon) = 0, \quad y'(1,\varepsilon) = 1, \end{aligned} \tag{E_{15}}$$

whose unique solution is simply $y(t,\varepsilon) = t$. Nevertheless, $u = u_L(t) = -t$ and $u = u_R(t) = t - 1$ are stable solutions of the corresponding reduced problems which intersect at $t_0 = 1/2$ and satisfy the Haber-Levinson condition that

$$1 - \lambda^4 > 0 \quad \text{for} \quad u_L' = -1 < \lambda < 1 = u_R'.$$

Unfortunately, there is <u>no</u> solution of (E_{15}) which is close to the angular path $\max\{-t,t-1\}$ in $(0,1)$.

In order to formulate the correct result, we proceed as follows. First of all, let us assume that the reduced problem

$$\begin{aligned} &f(t,u,u') = 0, \\ &u(a) - p_1 u'(a) = A, \end{aligned} \tag{R_L}$$

and the reduced problem (R_R) have, respectively, solutions $u = u_L(t)$ in $[a,t_0]$ and $u = u_R(t)$ in $[t_0,b]$, where $a < t_0 < b$, $u_L(t_0) = u_R(t_0)$ $(= \sigma)$ and $\nu_L = u_L'(t_0) \neq u_R'(t_0) = \nu_R$.

Let

$$u(t) = \begin{cases} u_L(t), & a \le t \le t_0, \\ u_R(t), & t_0 \le t \le b. \end{cases}$$

We naturally want u to be stable in the sense of Definition 6.2, that is,

$$f_{y'}(t,y,y') \ge k > 0 \quad \text{in} \quad [a,t_0] \tag{6.6}$$

and

$$f_{y'}(t,y,y') \le -k < 0 \quad \text{in} \quad [t_0,b] \tag{6.7}$$

in the region $D_1(u)$. At the crossing point t_0, we must further require that

$$(\nu_R - \nu_L) f(t_0,\sigma,\lambda) > 0 \tag{6.8}$$

for all λ strictly between ν_L and ν_R. Now in view of Example (E_{15}), we must add a further condition, namely,

$$f_y(t,y,y') \ge m > 0 \tag{6.9}$$

in the region $D_1(u)$. With these assumptions, we can formulate the following result.

<u>Theorem 6.12</u>. Assume that the reduced problems (R_L) and (R_R) have solutions $u = u_L(t)$, $u = u_R(t)$ of class $C^{(2)}([a,t_0])$, $C^{(2)}([t_0,b])$, respectively, which satisfy the relations (6.6) - (6.9). Then there exists

an $\varepsilon_0 > 0$ such that for $0 < \varepsilon \leq \varepsilon_0$ the problem (RP_9) has a solution $y = y(t,\varepsilon)$ in $[a,b]$ satisfying

$$y(t,\varepsilon) = u(t) + O(\varepsilon|\nu_R-\nu_L|(2k)^{-1}\exp[-k|t-t_0|/\varepsilon]) + O(\varepsilon).$$

Proof: The proof proceeds in the now familiar manner by defining the appropriate bounding functions. It is also not difficult to see that any solution $y = y(t,\varepsilon)$ of $\varepsilon y'' = f(t,y,y')$ which lies between the bounding functions is such that

$$y'(t,\varepsilon) = u'(t) + O(|\nu_R-\nu_L|/2 \exp[-k|t-t_0|/\varepsilon]) + O(\varepsilon),$$

and so we can apply Theorem 2.3 to obtain the required result.

§6.7. A Comment

We conclude this chapter with the remark that under certain circumstances (cf. [44]) solutions of the Dirichlet problem (DP_5) exist and exhibit boundary and shock layer behavior. In particular, if

$$f(t,y,y') = p(t,y)y'^n + O(|y'|^{n-1}) \quad \text{as} \quad |y'| \to \infty,$$

and if p *vanishes* at a value of t in $[a,b]$ for all y or along a path (t,y) in $[a,b] \times \mathbb{R}^1$, then the result of Vishik and Liusternik prohibiting boundary layer behavior does not apply. As a simple illustration, consider the problem

$$\varepsilon y'' = -ty'^3 \equiv f(t,y'), \quad 0 < t < 1,$$
$$y(0,\varepsilon) = 0, \quad y(1,\varepsilon) = B > 0. \qquad (E_{16})$$

Its solution, which can be found by quadratures, satisfies

$$\lim_{\varepsilon\to 0+} y(t,\varepsilon) = B \quad \text{for} \quad 0 < \delta \leq t \leq 1,$$

that is, there is a boundary layer at $t = 0$. Note that $f(t,y') = O(|y'|^3)$ as $|y'| \to \infty$, for fixed t in $(0,1]$, but $f(0,y') \equiv 0$. The interested reader should consult [44] for a detailed discussion of such "superquadratic" problems.

Notes and Remarks

6.1. As in the other chapters we can allow the righthand side f and the boundary data (that is, p_1, p_2, A and B) to depend on ε in a regular manner.

6.2. The results here are due to Howes in a series of papers (cf. [40], [42], [44]). Recently Deshpande [19] has systematically considered the existence and nonexistence of solutions of the Dirichlet problem (6.1), (6.2) as $\varepsilon \to 0^+$ for certain classes of superquadratic functions f.

6.3. Superquadratic boundary value problems of the type just discussed arise in many applied contexts involving capillarity and/or surface tension effects (see, for example, [50], [78], [79] and Chapter VIII below). Studies of related problems on unbounded t-intervals can be found in [50], [65], [58] and [42].

Chapter VII
Singularly Perturbed Systems

§7.1. Introduction

In this chapter we turn our attention to some vector boundary value problems which may be regarded as vector analogs of the scalar problems. However, as the reader will see, our results for vector problems are very incomplete, especially in comparison with the scalar theory. The study of singularly perturbed vector second-order equations is in its infancy, and we hope that this chapter will serve to draw attention to some of the associated difficulties. Indeed, many fundamental questions concerning vector equations or systems have yet to be raised, let alone answered.

§7.2. The Semilinear Dirichlet Problem

Let us begin with the semilinear problem

$$\varepsilon \underset{\sim}{y}'' = \underset{\sim}{H}(t,\underset{\sim}{y}), \quad a < t < b,$$
$$\underset{\sim}{y}(a,\varepsilon) = \underset{\sim}{A}, \quad \underset{\sim}{y}(b,\varepsilon) = \underset{\sim}{B}, \qquad (S_1)$$

where $\underset{\sim}{y}$, $\underset{\sim}{A}$ and $\underset{\sim}{B}$ are N-vectors (interpreted throughout as column vectors), and $\underset{\sim}{H}$ is an N-vector function which is defined and smooth on $[a,b] \times \mathbb{R}^N$. The assumed smallness of $\varepsilon > 0$ prompts us, as usual, to look first for solutions of the reduced system

$$\underset{\sim}{H}(t,\underset{\sim}{u}) = \underset{\sim}{0}. \qquad (R_1)$$

Let us assume, for simplicity, that this reduced system has the trivial solution $\underset{\sim}{u} \equiv \underset{\sim}{0}$ in $[a,b]$. In order to get some feeling for the types of conditions on $\underset{\sim}{H}$ which guarantee the existence of solutions of (S_1)

that are close to zero almost everywhere in $[a,b]$, we first look at the simple linear system

$$\varepsilon \underset{\sim}{y}'' = A\underset{\sim}{y}, \quad a < t < b, \qquad (E_{17})$$
$$\underset{\sim}{y}(z,\varepsilon) = \underset{\sim}{A}, \quad \underset{\sim}{y}(b,\varepsilon) = \underset{\sim}{B},$$

where A is a constant, symmetric $(N \times N)$-matrix. If A is positive definite, then it is similar to a diagonal matrix with positive entries $m_1,\ldots,m_N$. Thus the scalar theory (cf. Example (E_1) in Chapter III) implies that, under this assumption, Example (E_{17}) has a solution $\underset{\sim}{y} = \underset{\sim}{y}(t,\varepsilon)$ for all $\varepsilon > 0$ which satisfies

$$\lim_{\varepsilon \to 0^+} \underset{\sim}{y}(t,\varepsilon) = \underset{\sim}{0} \quad \text{for } t \text{ in } [a+\delta, b-\delta], \tag{7.1}$$

where $0 < \delta < b-a$. Indeed, the reader should have no difficulty showing that each component of the solution $\underset{\sim}{y}(t,\varepsilon)$ satisfies the more precise estimate in $[a,b]$:

$$y_i(t,\varepsilon) = \mathcal{O}(|A_i|\exp[-(m_i/\varepsilon)^{1/2}(t-a)])$$
$$+ \mathcal{O}(|B_i|\exp[-(m_i/\varepsilon)^{1/2}(b-t)]),$$

for $i = 1,\ldots,N$. Generalizing to the system (S_1), we may anticipate that if there is a positive constant m such that $J_0(t)$, the Jacobian matrix of $\underset{\sim}{H}$ evaluated at $\underset{\sim}{y} = \underset{\sim}{0}$, satisfies the inequality

$$\underset{\sim}{y}^T J_0 \underset{\sim}{y} \geq m||\underset{\sim}{y}||^2 \quad \text{in } [a,b], \tag{7.2}$$

for all $\underset{\sim}{y}$ in $\mathbb{R}^N$, then the problem (S_1) will have a solution which satisfies the limiting relation (7.1), *provided* $||\underset{\sim}{A}||$ and $||\underset{\sim}{B}||$ are sufficiently small. This restriction (7.2) is precisely the condition that the coefficient matrix of the corresponding linearized system sould be positive definite.

To consolidate the above heuristic ideas, let us first define the following region in $\mathbb{R}^{N+1}$:

$$\mathcal{D} = \{(t,\underset{\sim}{y}): a \leq t \leq b, \ ||\underset{\sim}{y}|| \leq d(t)\}.$$

Here $d(t)$ is a smooth positive function such that $d(t) \equiv ||\underset{\sim}{A}|| + \delta$ in $[a, a + \delta/2]$, $d(t) \equiv ||\underset{\sim}{B}|| + \delta$ in $[b - \delta/2, b]$ and $d(t) \equiv \delta$ in $[a + \delta, b - \delta]$, where $\delta > 0$. The following definition of stability for the reduced solution of (R_1) may be regarded as the vector analog of Definition 3.1.

Definition 7.1. The zero solution $\underset{\sim}{u} \equiv \underset{\sim}{0}$ of the reduced system (R_1) is said to be norm-stable (in $[a,b]$) if there is a positive constant m such that

$$\underset{\sim}{y}^T J(t,\underset{\sim}{y})\underset{\sim}{y} \geq m||\underset{\sim}{y}||^2 \quad \text{for} \quad (t,\underset{\sim}{y}) \quad \text{in} \quad \mathcal{D}, \tag{7.3}$$

where $J(t,\underset{\sim}{y}) \equiv (\partial \underset{\sim}{H}/\partial \underset{\sim}{y})(t,\underset{\sim}{y})$ is the Jacobian matrix of $\underset{\sim}{H}$ with respect to $\underset{\sim}{y}$.

The condition (7.3) is obviously stronger than the condition (7.2). We will see that if $\underset{\sim}{u} \equiv \underset{\sim}{0}$ is norm-stable, then for all sufficiently small $\varepsilon > 0$ there exists a solution of the semilinear problem (S_1) satisfying the limiting relation (7.1). This is the content of the first theorem (cf. [54], [41]).

Theorem 7.1. Assume that the zero function is a norm-stable solution of the reduced system (R_1) in $[a,b]$. Then there exists an $\varepsilon_0 > 0$ such that for $0 < \varepsilon \leq \varepsilon_0$ the problem (S_1) has a solution $\underset{\sim}{y} = \underset{\sim}{y}(t,\varepsilon)$ of class $C^{(2)}([a,b])$ satisfying

$$||\underset{\sim}{y}(t,\varepsilon)|| \leq ||\underset{\sim}{A}||\exp[-(m\varepsilon^{-1})^{1/2}(t-a)] + ||\underset{\sim}{B}||\exp[-(m\varepsilon^{-1})^{1/2}(b-t)]. \tag{7.4}$$

Proof: The result is a consequence of Theorem 2.4 if we can construct a scalar function

$$\rho = \rho(t,\underset{\sim}{y},\varepsilon) = ||\underset{\sim}{y}|| - \gamma(t,\varepsilon)$$

with ρ satisfying (2.5), or equivalently with γ satisfying

$$\varepsilon\gamma'' \leq \underset{\sim}{y}^T \underset{\sim}{H}(t,\underset{\sim}{y})/||\underset{\sim}{y}|| \quad \text{whenever} \quad \gamma = ||\underset{\sim}{y}||.$$

(Refer to the discussions immediately following Theorem 2.4).

Define

$$\gamma = \gamma(t,\varepsilon) = ||\underset{\sim}{A}||\exp[-(m\varepsilon^{-1})^{1/2}(t-a)] + ||\underset{\sim}{B}||\exp[-(m\varepsilon^{-1})^{1/2}(b-t)].$$

Then γ satisfies $\varepsilon\gamma'' = m\gamma$. Note that the set $(t,\underset{\sim}{y})$, with $a \leq t \leq b$, $||\underset{\sim}{y}|| \leq \gamma$, is in $\mathcal{D}$ for all suffficiently small ε, say $0 < \varepsilon \leq \varepsilon_0$. From the Mean Value Theorem and (7.3) it follows that for $(t,\underset{\sim}{y})$ in $\mathcal{D}$,

$$\underset{\sim}{y}^T \underset{\sim}{H}(t,\underset{\sim}{y})/||\underset{\sim}{y}|| = \underset{\sim}{y}^T J(t,\underset{\sim}{\xi})\underset{\sim}{y}/||\underset{\sim}{y}|| \geq m||\underset{\sim}{y}||,$$

where $(t,\underset{\sim}{\xi})$ is the intermediate point, and so, whenever $\gamma = ||\underset{\sim}{y}||$ we have

$$\underset{\sim}{y}^T \underset{\sim}{H}(t,\underset{\sim}{y})/||\underset{\sim}{y}|| - \varepsilon\gamma'' \geq m\gamma - \varepsilon\gamma'' = 0.$$

We conclude from Theorem 2.4 that for $0 < \varepsilon \leq \varepsilon_0$, the problem (S_1) has a solution $\underset{\sim}{y} = \underset{\sim}{y}(t,\varepsilon)$ of class $C^{(2)}([a,b])$ such that $\rho(t,\underset{\sim}{y}(t,\varepsilon),\varepsilon) \leq 0$, that is, $||\underset{\sim}{y}(t,\varepsilon)|| \leq \gamma(t,\varepsilon)$.

It is possible to prove a more general result than Theorem 7.1, in which a limiting relation of the form (7.4) is still valid (cf. [41]). We simply replace the inequality (7.3) by the condition that there exist a smooth, real-valued function $h = h(t,z)$ such that, for $(t,\underset{\sim}{y})$ in $\mathcal{D}$,

$$(\underset{\sim}{y}^T/||\underset{\sim}{y}||)\underset{\sim}{H}(t,\underset{\sim}{y}) \geq h(t,||\underset{\sim}{y}||),$$

and h satisfies:

$$h(t,0) = 0; \quad \frac{\partial h}{\partial z}(t,0) \geq m > 0, \quad \text{for} \quad a \leq t \leq b;$$

$$\int_0^\xi h(a,s)ds > 0 \quad \text{for} \quad 0 < \xi \leq ||\underset{\sim}{A}||,$$

and

$$\int_0^\eta h(b,s)ds > 0 \quad \text{for} \quad 0 < \eta \leq ||\underset{\sim}{B}||.$$

(See Remark 3.3 at the end of Chapter III and Example 8.21 in Chapter VIII.)

Thus far we have concentrated on the bounds for the norm of the solution of (S_1), and in so doing, we have ignored the behavior of individual components as $\varepsilon \to 0^+$. Let us now seek conditions on the righthand side $\underset{\sim}{H}$ which will allow the study of each component of the solution (cf. [68], [69]). We first define the N regions in $\mathbb{R}$:

$$\mathcal{D}_i = \{y_i: |y_i| \leq d_i(t)\},$$

where each d_i is a smooth positive function such that $d_i(t) \equiv |A_i| + \delta$ in $[a,a + \delta/2]$, $d_i(t) \equiv |B_i| + \delta$ in $[b - \delta/2,b]$, and $d_i(t) \equiv \delta$ in $[a + \delta, b - \delta]$. Then we assume that the reduced system $(\mathcal{R}_1)$ has a solution $\underset{\sim}{u} \equiv \underset{\sim}{0}$, not in the sense that $\underset{\sim}{H}(t,\underset{\sim}{0}) = \underset{\sim}{0}$, but in the much stronger sense that

$$H_i(t,y_1,\ldots,y_{i-1},0,y_{i+1},\ldots,y_N) = 0, \quad i = 1,\ldots,N, \tag{7.5}$$

for t in $[a,b]$ and for all y_j in $\mathcal{D}_j$ $(j \neq i)$. In other words, we are only studying the class of problems (S_1) such that (7.5) has a solution $y_i \equiv 0$. The assumption (7.5) has the effect of decoupling the vector second-order system into N second-order scalar equations, so that the methods of Chapter III can be applied to estimate each component.

Following Definition 3.1, we now make the following definition of stability for the reduced solution.

Definition 7.2. The zero solution of the reduced system (R_1) (in the sense of (7.5)) is said to be *componentwise-stable* if there exist positive constants m_i such that

$$(\partial H_i/\partial y_i)(t,y_1,\dots,y_N) \geq m_i > 0, \qquad 1 \leq i \leq N, \tag{7.6}$$

for t in $[a,b]$ and y_i in $\mathcal{D}_i$.

The condition (7.6) is clearly the system analog of the scalar condition (cf. Definition 3.1), and it allows us to establish the next theorem (cf. [68], [69]).

Theorem 7.2. Assume that the equations (7.5) have a componentwise-stable solution $u_i = 0$. Then there exists an $\varepsilon_0 > 0$ such that for $0 < \varepsilon \leq \varepsilon_0$ the problem (S_1) has a solution $\underset{\sim}{y} = \underset{\sim}{y}(t,\varepsilon)$ of class $C^{(2)}([a,b])$ satisfying

$$|y_i(t,\varepsilon)| \leq |A_i|\exp[-(m_i\varepsilon^{-1})^{1/2}(t-a)] + |B_i|\exp[-(m_i\varepsilon^{-1})^{1/2}(b-t)].$$

Proof: The proof of this result is very similar to the proof of Theorem 3.1, and will follow from Theorem 2.4. For definiteness, let us suppose that $A_i > 0$ and $B_i < 0$, for some i; the other cases can be handled analogously. For such A_i and B_i, we define the pairs of functions

$$\alpha_i(t,\varepsilon) = B_i \exp[-(m_i\varepsilon^{-1})^{1/2}(b-t)]$$

and

$$\beta_i(t,\varepsilon) = A_i \exp[-(m_i\varepsilon^{-1})^{1/2}(t-a)].$$

The functions α_i and β_i are solutions of $\varepsilon v_i'' = m_i v_i$. Clearly $\alpha_i \leq \beta_i$, $\alpha_i(a,\varepsilon) \leq A_i \leq \beta_i(a,\varepsilon)$ and $\alpha_i(b,\varepsilon) \leq B_i \leq \beta_i(b,\varepsilon)$; moreover, the region $[a,b] \times [\alpha_i(t,\varepsilon),\beta_i(t,\varepsilon)]$ is contained in the region $[a,b] \times \mathcal{D}_i$ (for fixed $\delta > 0$) when ε is sufficiently small. Similarly, for bounding functions α_j, β_j $(j \neq i)$ defined analogously the region $[a,b] \times [\alpha_j(t,\varepsilon),\beta_j(t,\varepsilon)]$ is contained in $[a,b] \times \mathcal{D}_j$, when ε is sufficiently small.

Finally we show that α_i satisfies the required differential inequality; the verification for β_i is similar and is omitted. It follows from the Mean Value Theorem that for y_j in $\mathcal{D}_j$ $(j \neq i)$

$$\varepsilon\alpha_i'' - H_i(t,y_1,\ldots,\alpha_i,\ldots,y_N)$$

$$= m_i\alpha_i - H_i(t,y_1,\ldots,0,\ldots,y_N)$$

$$- (\partial H_i/\partial y_i)(t,y_1,\ldots,\Theta\alpha_i,\ldots,y_N)\alpha_i,$$

where $0 < \Theta < 1$. Now since $\alpha_i \leq \Theta\alpha_i \leq \beta_i$, we know that $\Theta\alpha_i$ is in $\mathcal{D}_i$, and if $\alpha_j \leq y_j \leq \beta_j$ $(j \neq i)$, then we know that y_j is in $\mathcal{D}_j$, provided of course that ε is sufficiently small, say $0 < \varepsilon \leq \varepsilon_0$. Consequently we may apply (7.6) to conclude that

$$\varepsilon\alpha_i'' - H_i(t,y_1,\ldots,\alpha_i,\ldots,y_N) \geq m_i\alpha_i - m_i\alpha_i = 0,$$

as required. It follows from Theorem 2.4 that for $0 < \varepsilon \leq \varepsilon_0$ and $a \leq t \leq b$ the problem (S_1) has a solution $\underset{\sim}{y}(t,\varepsilon)$ satisfying $\alpha_i \leq y_i(t,\varepsilon) \leq \beta_i$, that is,

$$B_i \exp[-(m_i\varepsilon^{-1})^{1/2}(b-t)] \leq y_i(t,\varepsilon) \leq A_i \exp[-(m_i\varepsilon^{-1})^{1/2}(t-a)].$$

As was the case with Theorem 7.1, this result can be improved if certain integral conditions are used. We refer the interested reader to O'Donnell [69] and also to Example 8.22 in Chapter VIII.

§7.3. The Semilinear Robin Problem

It is possible to obtain similar results for the following Robin problem

$$\begin{aligned} &\varepsilon\underset{\sim}{y}'' = \underset{\sim}{H}(t,\underset{\sim}{y}), \quad a < t < b, \\ &P\underset{\sim}{y}(a,\varepsilon) - \underset{\sim}{y}'(a,\varepsilon) = \underset{\sim}{A}, \quad Q\underset{\sim}{y}(b,\varepsilon) + \underset{\sim}{y}'(b,\varepsilon) = \underset{\sim}{B}, \end{aligned} \tag{S_2}$$

where P and Q are constant (N×N)-matrices. By analogy with the scalar Robin problem, we assume that P and Q are positive semidefinite in the sense that there exist nonnegative scalars p and q such that

$$\underset{\sim}{y}^T P\underset{\sim}{y} \geq p||\underset{\sim}{y}||^2, \quad \underset{\sim}{y}^T Q\underset{\sim}{y} \geq q||\underset{\sim}{y}||^2,$$

for any $\underset{\sim}{y}$ in $\mathbb{R}^N$. If we assume, for simplicity, that $\underset{\sim}{H}(t,\underset{\sim}{0}) \equiv \underset{\sim}{0}$ in $[a,b]$, then we might ask under what conditions on $\underset{\sim}{H}$ will the problem (S_2) have a solution which is close to the zero vector in $[a,b]$. The results just found for the Dirichlet problem (S_1) suggest that one sufficient condition is the existence of a *scalar* function $h = h(t,z)$ with the following properties:

$$\left.\begin{array}{l} h(t,0) \equiv 0 \\ (\partial h/\partial z)(t,0) \geq m > 0 \end{array}\right\} \quad \text{for } t \text{ in } [a,b]; \tag{7.7}$$

$$\underset{\sim}{y}^T \underset{\sim}{H}(t,\underset{\sim}{y})/||\underset{\sim}{y}|| \geq h(t,||\underset{\sim}{y}||), \tag{7.8}$$

for $(t,\underset{\sim}{y})$ in the region $\tilde{\mathcal{D}} \equiv [a,b] \times \{\underset{\sim}{y}$ in $\mathbb{R}^N$: $||\underset{\sim}{y}|| \leq \delta\}$, where δ is a small positive constant. Now we may recall from the results of Chapter III that under the condition (7.7), the Robin problem for $\varepsilon z'' = h(t,z)$ has a nonnegative solution $z(t,\varepsilon)$ which is asymptotically zero in $[a,b]$. Consequently, the *norm* of a solution $\underset{\sim}{y}(t,\varepsilon)$ of (S_2) will also be close to zero, if we can show that $||\underset{\sim}{y}(t,\varepsilon)|| \leq z(t,\varepsilon)$, as a result of the inequality (7.8). The precise statement is contained in the following theorem.

Theorem 7.3. Assume that the reduced system (R_1) has the solution $\underset{\sim}{u} \equiv \underset{\sim}{0}$ in $[a,b]$ and that there exists a scalar function $h = h(t,z)$, continuous with respect to t, z and continuously differentiable with respect to z, for (t,z) in $[a,b] \times \{z: |z| \leq \delta\}$, satisfying (7.7) and (7.8). Then there exists an $\varepsilon_0 > 0$ such that for $0 < \varepsilon \leq \varepsilon_0$ the problem (S_2) has a solution $\underset{\sim}{y} = \underset{\sim}{y}(t,\varepsilon)$ satisfying

$$||\underset{\sim}{y}(t,\varepsilon)|| \leq (\varepsilon m^{-1})^{1/2}||\underset{\sim}{A}||\exp[-(m\varepsilon^{-1})^{1/2}(t-a)]$$
$$+ (\varepsilon m^{-1})^{1/2}||\underset{\sim}{B}||\exp[-(m\varepsilon^{-1})^{1/2}(b-t)].$$

Proof: The proof is an easy application of Theorem 2.5. For t in $[a,b]$ and $\varepsilon > 0$, define the scalar function $\rho(t,\underset{\sim}{y}) = ||\underset{\sim}{y}|| - z(t,\varepsilon)$, where $z = z(t,\varepsilon)$ is the nonnegative solution of the scalar problem in (a,b):

$$\varepsilon z'' = h(t,z), \quad pz(a,\varepsilon) - z'(a,\varepsilon) = ||\underset{\sim}{A}||, \quad qz(b,\varepsilon) + z'(b,\varepsilon) = ||\underset{\sim}{B}||.$$

Now,

$$\begin{aligned} p\rho(a) - \rho'(a) &= p||\underset{\sim}{y}(a)|| - pz(a) - (\underset{\sim}{y}^T(a)/||\underset{\sim}{y}(a)||)\underset{\sim}{y}'(a) + z'(a) \\ &= p||\underset{\sim}{y}(a)|| - pz(a) + (\underset{\sim}{y}^T(a)/||\underset{\sim}{y}(a)||)\underset{\sim}{A} \\ &\quad - \underset{\sim}{y}^T(a)P(\underset{\sim}{y}(a)/||\underset{\sim}{y}(a)||) + z'(a) \\ &\leq p||\underset{\sim}{y}(a)|| - pz(a) + (\underset{\sim}{y}^T(a)/||\underset{\sim}{y}(a)||)\underset{\sim}{A} \\ &\quad - p||\underset{\sim}{y}(a)|| + z'(a) \end{aligned}$$

$$= -||\underset{\sim}{A}|| + (\underset{\sim}{y}^T(a)/||\underset{\sim}{y}(a)||)\underset{\sim}{A}$$
$$\leq 0,$$

since $-\underset{\sim}{y}^T(a)P\underset{\sim}{y}(a) \leq p||\underset{\sim}{y}(a)||^2$ by the definiteness of P, and $(\underset{\sim}{y}^T(a)/||\underset{\sim}{y}(a)||)\underset{\sim}{A} \leq ||\underset{\sim}{A}||$ by the Cauchy-Schwarz inequality. Similarly, using the definiteness of Q, it follows that $q\rho(b) + \rho'(b) \leq 0$. Finally we show that $\varepsilon z'' \leq (\underset{\sim}{y}^T/||\underset{\sim}{y}||)\underset{\sim}{H}(t,\underset{\sim}{y})$ whenever $z = ||\underset{\sim}{y}||$; indeed,

$$(\underset{\sim}{y}^T/||\underset{\sim}{y}||)\underset{\sim}{H}(t,\underset{\sim}{y}) - \varepsilon z''$$
$$\geq h(t,||\underset{\sim}{y}||) - \varepsilon z''$$
$$= h(t,z) - \varepsilon z''$$
$$= 0,$$

by our choice of the function z, provided ε is sufficiently small, say $0 < \varepsilon \leq \varepsilon_0$. Thus Theorem 2.5 tells us that the problem (S_2) has a smooth solution $\underset{\sim}{y} = \underset{\sim}{y}(t,\varepsilon)$ satisfying $\rho(t,\underset{\sim}{y}(t,\varepsilon)) \leq 0$ in $[a,b]$, that is,

$$||\underset{\sim}{y}(t,\varepsilon)|| \leq z(t,\varepsilon) \leq (\varepsilon m^{-1})^{1/2}||\underset{\sim}{A}||\exp[-(m\varepsilon^{-1})^{1/2}(t-a)]$$
$$+ (\varepsilon m^{-1})^{1/2}||\underset{\sim}{B}||\exp[-(m\varepsilon^{-1})^{1/2}(b-t)],$$

by virtue of Theorem 3.4.

We note that if the reduced system $\underset{\sim}{H}(t,\underset{\sim}{u}) = \underset{\sim}{0}$ has a smooth, nonzero solution $\underset{\sim}{u} = \underset{\sim}{u}(t)$, and if the inequalities (7.3) and (7.7), (7.8) are modified accordingly, then results analogous to Theorems 7.1 and 7.3 are valid. These modified conditions allow us to derive an estimate for $||\underset{\sim}{y}(t,\varepsilon) - \underset{\sim}{u}(t)||$ in $[a,b]$ as $\varepsilon \to 0^+$. It is also possible to obtain componentwise bounds on solutions of the Robin problem (S_2), in much the same way we obtained bounds in Theorem 7.2 on solutions of the Dirichlet problem (S_1). Indeed, for the Robin problem (S_2) the assumptions (7.5) and (7.6) reduce to the simpler assumptions that for t in $[a,b]$ and $i = 1,\ldots,N$

$$H_i(t,y_1,\ldots,y_{i-1},0,y_{i+1},\ldots,y_N) \equiv 0 \tag{7.5a}$$

and

$$(\partial H_i/\partial y_i)(t,y_1,\ldots,y_N) \geq m_i > 0, \quad \text{for } |y_i| \leq \delta. \tag{7.6a}$$

That is to say, we need only verify (7.6a) along the zero solution. With these ideas, we leave it to the interested reader to formulate for (S_2) a result which is analogous to Theorem 7.2.

§7.4. The Quasilinear Dirichlet Problem

We turn finally to an examination of the existence and the asymptotic behavior of solutions of the quasilinear vector problem

$$\varepsilon \underset{\sim}{y}'' = F(t,\underset{\sim}{y})\underset{\sim}{y}' + \underset{\sim}{g}(t,\underset{\sim}{y}), \quad a < t < b, \qquad (S_3)$$
$$\underset{\sim}{y}(a,\varepsilon) = \underset{\sim}{A}, \quad \underset{\sim}{y}(b,\varepsilon) = \underset{\sim}{B}.$$

Here F is a continuous $(N\times N)$-matrix-valued function and $\underset{\sim}{g}$ is a continuous N-vector-valued function, and each is continuously differentiable in $\underset{\sim}{y}$, on $[a,b] \times \mathbb{R}^N$. Depending on the properties of F and $\underset{\sim}{g}$, solutions of (S_3) can exhibit a variety of asymptotic behavior as $\varepsilon \to 0^+$; indeed, as we have already noted in Chapter IV, the scalar form of (S_3) is already fairly complicated. The principal difficulty in studying the system (S_3) arises from the coupling of the first-order derivatives in the righthand side. It is perhaps not surprising then that we must treat this problem under some rather restrictive conditions on F. We will study (S_3) in the same manner as the semilinear problem (S_1), by first considering norm-bound estimates on its solutions, followed by component-wise estimates, in the spirit of O'Donnell's work [68]. In order to apply O'Donnell's techniques we must assume that F is a diagonal matrix, that is, we assume (S_3) is a weakly coupled system in which the derivative of the i-th component appears only in the i-th equation. Norm-bound results can also be obtained for systems which are not necessarily weakly coupled; however, the estimates on the norm are usually much cruder than the corresponding estimates on the individual components.

Motivated by the scalar theory of Chapter IV, let us begin by considering solutions of the two reduced problems in (a,b)

$$F(t,\underset{\sim}{u})\underset{\sim}{u}' + \underset{\sim}{g}(t,\underset{\sim}{u}) = \underset{\sim}{0}, \quad \underset{\sim}{u}(a) = \underset{\sim}{A} \qquad (R_L)$$

and

$$F(t,\underset{\sim}{u})\underset{\sim}{u}' + \underset{\sim}{g}(t,\underset{\sim}{u}) = \underset{\sim}{0}, \quad \underset{\sim}{u}(b) = \underset{\sim}{B}, \qquad (R_R)$$

which are stable in the sense of Definitions 7.3, 7.4 respectively. We first define the regions

$$\mathcal{D}(\underset{\sim}{u}_L) = \{(t,\underset{\sim}{y})\colon a \le t \le b, \ ||\underset{\sim}{y}-\underset{\sim}{u}_L(t)|| \le d_L(t)\}$$

and

$$\mathcal{D}(\underset{\sim}{u}_R) = \{(t,\underset{\sim}{y})\colon a \le t \le b, \ ||\underset{\sim}{y}-\underset{\sim}{u}_R(t)|| \le d_R(t)\}.$$

Here d_L is a smooth positive function such that $d_L(t) \equiv ||\underset{\sim}{B}-\underset{\sim}{u}_L(b)|| + \delta$

for t in $[b-\delta/2,b]$ and $d_L(t) \equiv \delta$ for t in $[a,b-\delta]$, while d_R is a smooth positive function such that $d_R(t) \equiv ||\underset{\sim}{A}-\underset{\sim}{u}_R(a)|| + \delta$ for t in $[a,a+\delta/2]$ and $d_R(t) \equiv \delta$ for t in $[a+\delta,b]$.

<u>Definition 7.3</u>. A solution $\underset{\sim}{u} = \underset{\sim}{u}_L(t)$ of the reduced problem (R_L) is said to be *norm-stable* if there exists a positive constant k such that for $(t,\underset{\sim}{y})$ in $\mathcal{D}(\underset{\sim}{u}_L)$

$$\underset{\sim}{z}^T F(t,\underset{\sim}{y})\underset{\sim}{w} \geq k\underset{\sim}{z}^T\underset{\sim}{w},$$

for all $\underset{\sim}{z}$, $\underset{\sim}{w}$ in $\mathbb{R}^N$.

<u>Definition 7.4</u>. A solution $\underset{\sim}{u} = \underset{\sim}{u}_R(t)$ of the reduced problem (R_R) is said to be *norm-stable* if there exists a positive constant k such that for $(t,\underset{\sim}{y})$ in $\mathcal{D}(\underset{\sim}{u}_R)$

$$\underset{\sim}{z}^T F(t,\underset{\sim}{y})\underset{\sim}{w} \leq -k\underset{\sim}{z}^T\underset{\sim}{w},$$

for all $\underset{\sim}{z}$, $\underset{\sim}{w}$ in $\mathbb{R}^N$.

These definitions are obvious extensions of Definitions 4.1 and 4.2 for the scalar analog of (S_3). They imply that the matrix F is positive definite (negative definite) along $\underset{\sim}{u}_L$ $(\underset{\sim}{u}_R)$ and within the boundary layer at $t = b$ $(t = a)$. This is a rather strong restriction on F; however, this can be slightly weakened by means of certain integral conditions (cf. [47] and Example 8.24 in the next chapter). We can now state a basic result for the quasilinear problem (S_3).

<u>Theorem 7.4</u>. Assume that the reduced problem (R_L) has a norm-stable solution $\underset{\sim}{u} = \underset{\sim}{u}_L(t)$ of class $C^{(2)}([a,b])$. Then there exists an $\varepsilon_0 > 0$ such that for $0 < \varepsilon \leq \varepsilon_0$ the problem (S_3) has a solution $\underset{\sim}{y} = \underset{\sim}{y}(t,\varepsilon)$ satisfying

$$||\underset{\sim}{y}(t,\varepsilon) - \underset{\sim}{u}_L(t)|| \leq ||\underset{\sim}{B} - \underset{\sim}{u}_L(b)||\exp[-k_1\varepsilon^{-1}(b-t)] + K\varepsilon,$$

where K is a known positive constant and $0 < k_1 < k$.

<u>Proof</u>: In order to simplify the proof, let us introduce the new dependent variable $\underset{\sim}{v} = \underset{\sim}{y}-\underset{\sim}{u}_L(t)$, in terms of which the problem (S_3) becomes the problem

$$\begin{aligned} &\varepsilon\underset{\sim}{v}'' = \underset{\sim}{f}(t,\underset{\sim}{v},\underset{\sim}{v}',\varepsilon), \\ &\underset{\sim}{v}(a,\varepsilon) = \underset{\sim}{0}, \quad \underset{\sim}{v}(b,\varepsilon) = \underset{\sim}{B} - \underset{\sim}{u}_L(b), \end{aligned} \qquad (S_3')$$

where

$$\underline{f}(t,\underline{v},\underline{v}',\epsilon) = F(t,\underline{v} + \underline{u}_L(t))\underline{v}'$$

$$+ F(t,\underline{v} + \underline{u}_L(t))\underline{u}_L'(t) + g(t,\underline{v} + \underline{u}_L(t)) - \epsilon\underline{u}_L''(t).$$

The antitipcated application of Theorem 2.4 prompts us to define, for $a \le t \le b$ and $0 < \epsilon \le k^2/(4\ell)$, the function

$$\rho(t,\underline{v}) = ||\underline{v}|| - ||\underline{B} - \underline{u}_L(b)||\exp[\lambda_1(b-t)] - \epsilon\nu\ell^{-1}(\exp[\lambda_2(a-t)] - 1).$$

Here (cf. the proof of Theorem 4.1), $\lambda_1 \sim -k\epsilon^{-1}$ and $\lambda_2 \sim -\ell k^{-1}$ are the roots of the quadratic $\epsilon\lambda^2 + k\lambda + \ell$ (k is the positive constant in Definition 7.3, and ℓ is a positive constant such that $|||(\partial f/\partial v)(t,\underline{\eta},\underline{0},0)||| \le \ell$, for $(t,\underline{\eta})$ in $[a,b] \times \{\underline{\eta}: ||\underline{\eta}|| \le d_L(t)\}$; the matrix norm $|||\cdot|||$ is defined by $|||G|||^2 \equiv \sup||G\underline{z}||: ||\underline{z}|| = 1\})$. The positive constant ν will be determined later.

If we assume that

$$\gamma = ||\underline{v}|| \quad \text{and} \quad \gamma' = \underline{v}^T\underline{v}'/||\underline{v}||,$$

for

$$\gamma(t,\epsilon) = ||\underline{B} - \underline{u}_L(b)||\exp[\lambda_1(b-t)] + \epsilon\nu\ell^{-1}(\exp[\lambda_2(a-t)] - 1),$$

then Theorem 2.4 is applicable, if we can show that

$$(\underline{v}^T/||\underline{v}||)\underline{f}(t,\underline{v},\underline{v}',\epsilon) - \epsilon\gamma'' \ge 0 \quad (\text{in} \quad (a,b)).$$

Expanding by the Mean Value Theorem gives us

$$(\underline{v}^T/||\underline{v}||)\underline{f} - \epsilon\gamma''$$

$$= (\underline{v}^T/||\underline{v}||)[\underline{f}(t,\underline{0},\underline{0},0) + F(t,\underline{v}+\underline{u}_L(t))\underline{v}'$$

$$+ (\partial f/\partial v)(t,\underline{\eta},\underline{0},0)\underline{v} - \epsilon\underline{u}_L''(t)] - \epsilon\gamma''$$

$$\ge (\underline{v}^T/||\underline{v}||)F(t,\underline{v} + \underline{u}_L(t))\underline{v}'$$

$$+ (\underline{v}^T/||\underline{v}||)(\partial f/\partial v)(t,\underline{\eta},\underline{0},0)\underline{v}$$

$$- \epsilon L - \epsilon\gamma'',$$

since $\underline{f}(t,\underline{0},\underline{0},0) \equiv 0$. Here $(t,\underline{\eta},\underline{0},0)$ is the appropriate intermediate point and $L \equiv \max_{[a,b]} ||\underline{u}_L''(t)||$. Now if ϵ is sufficiently small, say $0 < \epsilon \le \epsilon_0$, then the point $(t,\underline{v} + \underline{u}_L(t))$ belongs to the region $\mathcal{D}(\underline{u}_L)$, and so the norm-stability of $\underline{u}_L$ allows us to continue with the inequality

$$\begin{aligned}
(\underset{\sim}{v}^T/||\underset{\sim}{v}||)\underset{\sim}{f} - \varepsilon\gamma'' \\
&\geq k\underset{\sim}{v}^T\underset{\sim}{v}'/||\underset{\sim}{v}|| - \ell||\underset{\sim}{v}|| - \varepsilon L - \varepsilon\gamma'' \\
&= -k\lambda_1||\underset{\sim}{B} - \underset{\sim}{u}_L(b)||\exp[\lambda_1(b-t)] \\
&\quad -k\lambda_2\varepsilon\nu\ell^{-1}\exp[\lambda_2(a-t)] \\
&\quad -\ell||\underset{\sim}{B} - \underset{\sim}{u}_L(b)||\exp[\lambda_1(b-t)] \\
&\quad -\ell\varepsilon\nu\ell^{-1}\exp[\lambda_2(a-t)] + \varepsilon\nu - \varepsilon L \\
&\quad - \varepsilon\lambda_1^2||\underset{\sim}{B} - \underset{\sim}{u}_L(b)||\exp[\lambda_1(b-t)] \\
&\quad - \varepsilon\lambda_2^2\varepsilon\nu\ell^{-1}\exp[\lambda_2(a-t)] \\
&= 0,
\end{aligned}$$

if we set $\nu = L$, since $\varepsilon\lambda_i^2 + k\lambda_i + \ell = 0$, $i = 1,2$. Therefore it follows from Theorem 2.4 that the problem (S_3'), and hence the original problem (S_3) have, respectively, $C^{(2)}$-solutions $\underset{\sim}{v} = \underset{\sim}{v}(t,\varepsilon)$, $\underset{\sim}{y} = \underset{\sim}{y}(t,\varepsilon)$, such that $\rho(t,\underset{\sim}{v}(t,\varepsilon)) \leq 0$, that is,

$$||\underset{\sim}{v}(t,\varepsilon)|| = ||\underset{\sim}{y}(t,\varepsilon) - \underset{\sim}{u}_L(t)|| \leq ||\underset{\sim}{B} - \underset{\sim}{u}_L(b)||\exp[-k_1\varepsilon^{-1}(b-t)] + K\varepsilon \quad \text{in } [a,b],$$

for $0 < k_1 < k$ and $K = L\ell^{-1}(\exp[\lambda_2(a-b)] - 1)$.

The companion result for a boundary layer at $t = a$ follows easily from Theorem 7.4 and Definition 7.4 via the change of variable $t \to a+b-t$. We leave its precise formulation to the reader.

For classes of problems such as (S_3), it is often advantageous to seek componentwise-bounds, rather than norm-bounds, on the solutions. In order to accomplish this, let us assume in what follows that the matrix F is diagonal, say $F(t,\underset{\sim}{y}) \equiv \operatorname{diag}\{f_1(t,\underset{\sim}{y}),\ldots,f_N(t,\underset{\sim}{y})\}$ for smooth functions f_i. Then the system (S_3) can be written in component form as

$$\begin{aligned}
&\varepsilon y_i'' = f_i(t,\underset{\sim}{y})y_i' + g_i(t,\underset{\sim}{y}), \quad a < t < b, \\
&y_i(a,\varepsilon) = A_i, \quad y_i(b,\varepsilon) = B_i.
\end{aligned} \tag{S_4}$$

Since the righthand side of the i-th equation depends only on y_i' and does not depend on y_j' $(j \neq i)$, we say that the quasilinear system is weakly coupled.

Let us now look for solutions of (S_4) which exhibit boundary layer behavior at $t = a$; analogous results for boundary layer behavior at $t = b$ then follow in the usual manner. With our first assumption that the reduced problem (R_R) has a solution $\underset{\sim}{u} = \underset{\sim}{u}(t)$ of class $C^{(2)}([a,b])$, we define the regions

$$\mathcal{D}_i = \{y_i : |y_i - u_i(t)| \leq d_i(t)\}, \quad i = 1,\ldots,N,$$

where each d_i is a smooth positive function such that $d_i(t) \equiv |A_i - u_i(a)| + \delta$ for t in $[a,a+\delta/2]$ and $d_i(t) \equiv \delta$ for t in $[a+\delta,b]$, with δ a small positive constant. The second assumption is that $\underset{\sim}{u}$ additionally satisfies the reduced differential equation in the following *strong* sense, namely, for $i = 1,\ldots,N$

$$f_i(t,\underset{\sim}{y}_{ui})u_i' + g_i(t,\underset{\sim}{y}_{ui}) = 0, \tag{7.9}$$

for all $(t,\underset{\sim}{y}_{ui}) \equiv (t,y_1,\ldots,y_{i-1},u_i,y_{i+1},\ldots,y_N)$ with y_j in $\mathcal{D}_j$, $j \neq i$. A solution of the reduced equations (7.9) will be called a strong reduced solution, to distinguish it from the reduced solution of the reduced system (R_R). As was the case with the semilinear problem (S_1) (cf. (7.5)), the second assumption is precisely the condition which allows us to decouple the system, and thereby apply the scalar theory of Chapter IV to the problem (S_4). Lastly we require $\underset{\sim}{u}$ to be stable in the following sense.

Definition 7.5. A strong reduced solution $\underset{\sim}{u} = \underset{\sim}{u}(t)$ of the reduced problem (R_R) is said to be *componentwise-stable* (in $[a,b]$) if there are positive constants k_i such that

$$f_i(t,\underset{\sim}{y}) \leq -k_i < 0,$$

for all $(t,\underset{\sim}{y})$ in the region $\mathcal{D} \equiv [a,b] \times \prod_{i=1}^{N} \mathcal{D}_i$.

With this notion of stability, we have the following result (cf. [68]).

Theorem 7.5. Assume that the reduced problem (R_R) has a componentwise-stable strong solution $\underset{\sim}{u} = \underset{\sim}{u}(t)$ of class $C^{(2)}([a,b])$. Then there exists an $\varepsilon_0 > 0$ such that for $0 < \varepsilon \leq \varepsilon_0$ the problem (S_4) has a solution $\underset{\sim}{y} = \underset{\sim}{y}(t,\varepsilon)$ of class $C^{(2)}([a,b])$ satisfying, for $i = 1,\ldots,N$,

$$|y_i(t,\varepsilon) - u_i(t)| \leq |A_i - u_i(a)|\exp[-\tilde{k}_i\varepsilon^{-1}(t-a)] + K_i\varepsilon,$$

where $0 < \overline{k}_i < k_i$ and each K_i is a known positive constant.

Proof: It is enough to consider just the i-th component. We assume, for definiteness, that $u_i(a) \geq A_i$, and so we define the bounding functions

$$\alpha_i(t,\varepsilon) = u_i(t) - w_i(t,\varepsilon) - W_i(t,\varepsilon)$$

and

$$\beta_i(t,\varepsilon) = u_i(t) + W_i(t,\varepsilon),$$

where $w_i(t,\varepsilon) = (u_i(a)-A_i)\exp[\lambda_i(t-a)]$ and $W_i(t,\varepsilon) = \varepsilon\nu_i\ell_i^{-1}(\exp[\mu_i(t-b)]-1)$, in order to apply the "componentwise" version of Theorem 2.4. Here $\lambda_i \sim -k_i\varepsilon^{-1}$ and $\mu_i \sim -\ell_ik_i^{-1}$ are the negative roots of the quadratic $\varepsilon\lambda^2 + k_i\lambda + \ell_i$ (for $0 < \varepsilon \le k_i^2(4\ell_i)^{-1}$), ℓ_i is a positive constant such that

$$|(\partial f_i/\partial y_i)(t,\underset{\sim}{y})u_i' + (\partial g_i/\partial y_i)(t,\underset{\sim}{y})| \le \ell_i,$$

for $(t,\underset{\sim}{y})$ in $\mathcal{D}$, and $\nu_i \equiv \max_{[a,b]} |u_i''(t)|$. Clearly $\alpha_i \le \beta_i$, $\alpha_i(a,\varepsilon) \le A_i \le \beta_i(a,\varepsilon)$ and $\alpha_i(b,\varepsilon) \le B_i \le \beta_i(b,\varepsilon)$, and so it remains to verify that α_i, β_i satisfy the differential inequalities in (a,b):

$$\varepsilon\alpha_i'' \ge f_i(t,\underset{\sim}{y}_{\alpha i})\alpha_i' + g_i(t,\underset{\sim}{y}_{\alpha i}), \quad \varepsilon\beta_i'' \le f_i(t,\underset{\sim}{y}_{\beta i})\beta_i' + g_i(t,\underset{\sim}{y}_{\beta i}).$$

The vector $\underset{\sim}{y}_{vi}$ is equal to $(y_1,\ldots,y_{i-1},v_i,y_{i+1},\ldots,y_N)$, where $\alpha_j \le y_j \le \beta_j$ for appropriate bounding functions α_j, β_j, $j \ne i$, that is, $y_j = u_j(t) + O(|A_j-u_j(a)|\exp[-k_j\varepsilon^{-1}(t-a)] + O(\varepsilon)$, and so y_j is in $\mathcal{D}_j$ for ε sufficiently small, say $0 < \varepsilon \le \varepsilon_0$. We only verify that α_i satisfies the required inequality, as the verification for β_i proceeds analogously. Differentiating and expanding via the Mean Value Theorem, we have

$$\begin{aligned}
&\varepsilon\alpha_i'' - f_i(t,\underset{\sim}{y}_{\alpha i})\alpha_i' - g_i(t,\underset{\sim}{y}_{\alpha i}) \\
&\qquad = \varepsilon u_i'' - \varepsilon\lambda_i^2 w_i - \varepsilon\mu_i^2(W_i + \varepsilon\nu_i\ell_i^{-1}) \\
&\qquad\quad - f_i(t,\underset{\sim}{y}_{ui})u_i' - g_i(t,\underset{\sim}{y}_{ui}) \\
&\qquad\quad + [(\partial f_i/\partial y_i)(t,\underset{\sim}{\eta}_{\alpha i})u_i' + (\partial g_i/\partial y_i)(t,\underset{\sim}{\eta}_{\alpha i})](w_i + W_i) \\
&\qquad\quad + f_i(t,\underset{\sim}{y}_{\alpha i})[\lambda_i w_i + \mu_i(W_i + \varepsilon\nu_i\ell_i^{-1})],
\end{aligned}$$

where $\underset{\sim}{\eta}_{\alpha i} = (y_1,\ldots,y_{i-1},u_i + \Theta(\alpha_i-u_i),y_{i+1},\ldots,y_N)$, $0 < \Theta < 1$, is the appropriate intermediate point. Since, by assumption, $f_i(t,\underset{\sim}{y}_{ui})u_i' + g_i(t,\underset{\sim}{y}_{ui}) \equiv 0$ and $f_i(t,\underset{\sim}{y}_{\alpha i}) \le -k_i < 0$, we can continue with the inequality

$$\begin{aligned}
&\varepsilon\alpha_i'' - f_i(t,\underset{\sim}{y}_{\alpha i})\alpha_i' - g_i(t,\underset{\sim}{y}_{\alpha i}) \\
&\quad \geq -\varepsilon\nu_i - \varepsilon\lambda_i^2 w_i - \varepsilon\mu_i^2(W_i + \varepsilon\nu_i\ell_i^{-1}) \\
&\qquad - \ell_i w_i - \ell_i(W_i + \varepsilon\nu_i\ell_i^{-1}) + \varepsilon\nu_i - k_i\lambda_i w_i \\
&\qquad - k_i\mu_i(W_i + \varepsilon\nu_i\ell_i^{-1}) \\
&\quad = 0,
\end{aligned}$$

owing to the fact that $\varepsilon\lambda_i^2 + k_i\lambda_i + \ell_i = 0$ and $\varepsilon\mu_i^2 + k_i\mu_i + \ell_i = 0$.

We conclude from Theorem 2.4 that for $0 < \varepsilon \leq \varepsilon_0$ the problem (S_4) has a solution $\underset{\sim}{y} = \underset{\sim}{y}(t,\varepsilon)$ of class $C^{(2)}([a,b])$ satisfying $\alpha_i(t,\varepsilon) \leq y_i(t,\varepsilon) \leq \beta_i(t,\varepsilon)$ in $[a,b]$, that is,

$$-(u_i(a) - A_i)\exp[\lambda_i(t-a)] - K_i\varepsilon \leq y_i(t,\varepsilon) - u_i(t) \leq K_i\varepsilon,$$

for $K_i = \nu_i\ell_i^{-1}(\exp[\mu_i(a-b)] - 1)$.

The above result can, of course, be improved by appealing to componentwise integral conditions of the type mentioned at the end of Chapter IV (cf. [68] and Example 8.25). As noted before, the complementary theorem involving a solution of the reduced problem (R_L) and a boundary layer at $t = b$ follows from Theorem 7.5 by making the usual change of variable $t \to a + b - t$.

Finally it is possible to combine these two results into a "hybrid" theorem which can be proved in exactly the same manner as Theorem 7.5. It involves a solution $\underset{\sim}{u} = \underset{\sim}{U}(t)$ of the reduced problem

$$\begin{aligned}
&F(t,\underset{\sim}{u})\underset{\sim}{u}' + \underset{\sim}{g}(t,\underset{\sim}{u}) = \underset{\sim}{0}, \quad a < t < b, \\
&u_i(a) = A_i \quad (1 \leq i \leq M), \quad u_i(b) = B_i \quad (M+1 \leq i \leq N),
\end{aligned} \qquad (R_2)$$

as well as the regions

$$\tilde{\mathcal{D}}_i = \{ y_i : |y_i - U_i(t)| \leq \tilde{d}_i(t)\},$$

where $\tilde{d}_i$ is a smooth positive function such that $\tilde{d}_i(t) \equiv |B_i - U_i(b)| + \delta$ in $[b-\delta/2,b]$ and $\tilde{d}_i(t) \equiv \delta$ in $[a,b-\delta]$ for $i = 1,\ldots,M$, and $\tilde{d}_i(t) \equiv |A_i - U_i(a)| + \delta$ in $[a,a+\delta/2]$ and $\tilde{d}_i(t) \equiv \delta$ in $[a+\delta,b]$ for $i = M+1,\ldots,N$. A solution $\underset{\sim}{U}$ of (R_2) is then a strong solution if, in addition, it satisfies the system $(1 \leq i \leq N)$ on (a,b)

$$f_i(t,\underset{\sim}{y}_{Ui})U_i' + g_i(t,\underset{\sim}{y}_{Ui}) = 0,$$

for all $(t,\underset{\sim}{y}_{Ui}) \equiv (t,y_1,\ldots,y_{i-1},U_i,y_{i+1},\ldots,y_N)$ with y_j in $\tilde{\mathcal{D}}_j$, $j \neq i$.

This strong solution is said to be componentwise-stable if there are positive constants k_i $(1 \leq i \leq N)$ such that for all $(t,\tilde{y})$ in $[a,b] \times \prod_{i=1}^{N} \tilde{\mathcal{D}}_i$,

$$f_i(t,\tilde{y}) \geq k_i > 0 \quad \text{for} \quad i = 1,\dots,M$$

and

$$f_i(t,\tilde{y}) \leq -k_i < 0 \quad \text{for} \quad i = M+1,\dots,N.$$

Theorem 7.6. Assume that the reduced problem $(\mathcal{R}_2)$ has a componentwise-stable strong solution $\tilde{u} = \tilde{U}(t)$ of class $C^{(2)}([a,b])$. Then there exists an $\varepsilon_0 > 0$ such that for $0 < \varepsilon \leq \varepsilon_0$ the problem (S_4) has a solution $\tilde{y} = \tilde{y}(t,\varepsilon)$ of class $C^{(2)}([a,b])$ satisfying

$$|y_i(t,\varepsilon)-U_i(t)| \leq |B_i-U_i(b)|\exp[-\bar{k}_i\varepsilon^{-1}(b-t)] + K_i\varepsilon \quad (1 \leq i \leq M)$$

and

$$|y_i(t,\varepsilon)-U_i(t)| \leq |A_i-U_i(a)|\exp[-\bar{k}_i\varepsilon^{-1}(t-a)] + K_i\varepsilon \quad (M+1 \leq i \leq N),$$

where $0 < \bar{k}_i < k_i$ and K_i is a known positive constant.

Notes and Remarks

7.1. The theory of this chapter applies, with obvious modifications, to problems in which the righthand sides and boundary data depend regularly on ε.

7.2. It is possible to extend the scalar theory of interior layer phenomena, discussed in Chapters III and IV, to the semilinear and quasilinear systems considered in this chapter, if the appropriate reduced paths are componentwise-stable. The interested reader can consult the papers of O'Donnell [68], [69] for details and many examples.

7.3. In our discussion of the semilinear problems (S_1) and (S_2) we assumed that either $h_z \geq m > 0$ (for h such that $(\tilde{y}^T/||\tilde{y}||)\tilde{H}(t,\tilde{y}) \geq h(t,||\tilde{y}||)$) or $(\partial H_i/\partial y_i) \geq m_i > 0$ for $i = 1,\dots,N$. Since these are "scalar" conditions, we can easily apply the theory of Chapter III on higher-order stability conditions (cf. Definitions 3.1-3.6) to (S_1) and (S_2). The papers [41] and [46] are relevant in this regard.

7.4. The conditions (7.5) and (7.9), which guarantee that a reduced solution is a "strong" solution, deserve a brief comment. It turns out that these conditions are not invariant, under even a linear change of variables. In other words, it may be possible to transform a system like (S_1) or (S_4), not originally having any reduced solution satisfying (7.5) or (7.9), respectively, into a new system for which these conditions obtain.

7.5. Theorem 7.1 is due originally to Kelley [54] (cf. also [41]), Theorems 7.2, 7.5 and 7.6 are due to O'Donnell [68], [69], Theorem 7.3 is due to Howes [46] and Theorem 7.4 is due originally to Chang [12], who used a "diagonalization" method of approach. Earlier work on related problems includes the papers of Levin and Levinson [61], Levin [59], [60], Harris [32], Hoppensteadt [37], Chang and Coppel [13], Howes and O'Malley [47], as well as the monograph of Vasil'eva and Butuzov [88]. Additional references may be found in the monographs of Wasow [93] and O'Malley [75], and in O'Malley's long survey article [73].

Chapter VIII
Examples and Applications

Part I - SCALAR PROBLEMS

§8.1. Examples of Semilinear Problems and Applications

Example 8.1. Consider the Dirichlet problem

$$\varepsilon y'' = (y - u(t))^{2q+1}, \quad -1 < t < 1,$$
$$y(-1,\varepsilon) = A, \quad y(1,\varepsilon) = B,$$

where q is a nonnegative integer. If the function $u(t)$, defined for $-1 \leq t \leq 1$, is twice continuously differentiable or has a bounded second derivative, then by Theorem 3.1, for sufficiently small $\varepsilon > 0$, the Dirichlet problem has a solution $y = y(t,\varepsilon)$ which satisfies

$$\lim_{\varepsilon \to 0^+} y(t,\varepsilon) = u(t) \quad \text{in} \quad [-1+\delta, 1-\delta], \tag{8.1}$$

where $0 < \delta < 1$. Moreover, the behavior of the solution $y(t,\varepsilon)$ in the boundary layers at $t = -1$ and/or $t = 1$ (if $u(-1) \neq A$ and/or $u(1) \neq B$) can be described by means of the layer functions given in the conclusion of Theorem 3.1.

If we choose $u(t) = |t| = \max\{-t,t\}$, then the reduced solution is not differentiable at $t = 0$. In this situation the reduced solution u is best regarded as the union of the stable path $u_-(t) = -t$ in $[-1,0]$ and the stable path $u_+(t) = t$ in $[0,1]$. We can then apply Theorem 3.9 to deduce the existence of a solution $y = y(t,\varepsilon)$ which also satisfies the limiting relation (8.1). The precise behavior of the solution y in a neighborhood of the crossing point $t = 0$ of the reduced paths u_-, u_+ is also obtained from this theorem. Note that, as q becomes larger, the thickness of the angular layer at $t = 0$ increases correspondingly.

Example 8.2. Let us consider next the related problem

$$\varepsilon y'' = (y - u(t))^{2n}, \quad -1 < t < 1,$$

$$y(-1,\varepsilon) = A, \quad y(1,\varepsilon) = B,$$

where n is a positive integer and u is a convex function, say $u'' \geq 0$ in $(-1,1)$. The difference between this example and the last is that the right hand side is now raised to an even power. In this case, to ensure that there exists a solution $y(t,\varepsilon)$ satisfying the relation (8.1), we need to restrict the sign of both boundary layer 'jumps' $A - u(-1)$ and $B - u(1)$; indeed, we must require that $A \geq u(-1)$ and $B \geq u(1)$, as indicated in Theorem 3.2. If any of these inequalities is not satisfied, then our theory does not apply, and one can show that the Dirichlet problem has no solution of bounded t-variation as $\varepsilon \to 0^+$; cf. [76].

As an illustration, if $u(t) = t^2$ then for sufficiently small $\varepsilon > 0$, the problem has a solution $y = y(t,\varepsilon)$ satisfying $y(t,\varepsilon) \geq t^2$ and

$$\lim_{\varepsilon \to 0^+} y(t,\varepsilon) = t^2 \quad \text{in} \quad [-1+\delta, 1-\delta],$$

provided $A \geq 1$ and $B \geq 1$.

Example 8.3. As our third example we take the problem

$$\varepsilon y'' = y - y^3 \equiv h(y), \quad 0 < t < 1,$$

$$y(0,\varepsilon) = A, \quad y(1,\varepsilon) = B.$$

The reduced equation $h(u) = 0$ has three solutions $u_1 = 1$, $u_2 = 0$ and $u_3 = -1$, and since $h'(0) > 0$ while $h'(\pm 1) < 0$, we see that only u_2 is stable. Applying the integral condition (cf. Remark 3.3)

$$\int_0^{\xi} (s-s^3)\,ds > 0 \quad \text{provided} \quad 0 < |\xi| < \sqrt{2},$$

we find that if $|A| < \sqrt{2}$ and $|B| < \sqrt{2}$, then the problem has a solution as $\varepsilon \to 0^+$ such that

$$\lim_{\varepsilon \to 0^+} y(t,\varepsilon) = 0 \quad \text{in} \quad [\delta, 1-\delta],$$

by virtue of O'Malley's result [76].

The problem also has solutions exhibiting what is termed spike layer behavior, in that the solutions are asymptotically zero except at regularly spaced points. In a neighborhood of such a point the solution has a spike of finite height which does not vanish as $\varepsilon \to 0^+$; cf. Figures 8.1, 8.2. This follows, again from a result of O'Malley [76], because $u_2 = 0$ is a maximum point of the potential energy functional $\Psi(y) = -\int_A^y (s-s^3)ds = (y^4-A^4)/4 - (y^2-A^2)/2$ and $\Psi(\sqrt{2}) = \Psi(0) > 0$ (if $|A| < \sqrt{2}$), with $\sqrt{2}$ not a maximum point of Ψ. O'Malley's result implies that for each integer $n \geq 2$ the problem (with $|A|, |B| < \sqrt{2}$) has *four* solutions $y = y(t,\varepsilon)$ as $\varepsilon \to 0^+$ satisfying

$$\lim_{\varepsilon \to 0^+} y(t,\varepsilon) = 0 \quad \text{in} \quad [\delta, 1-\delta],$$

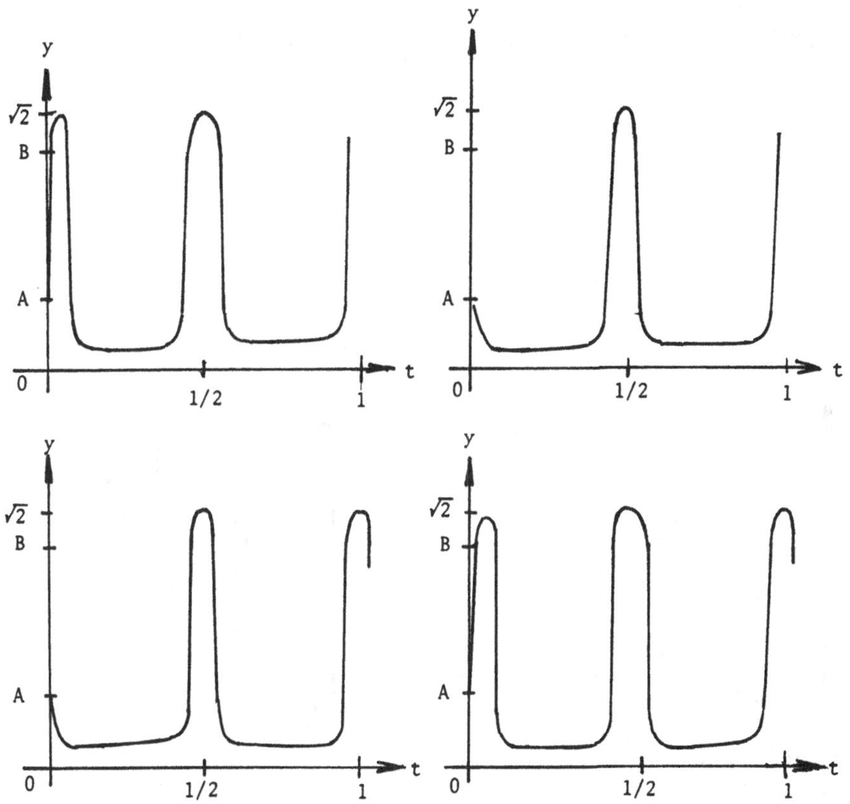

Figure 8.1
Spiked Solutions of Example 8.3 for n = 2.

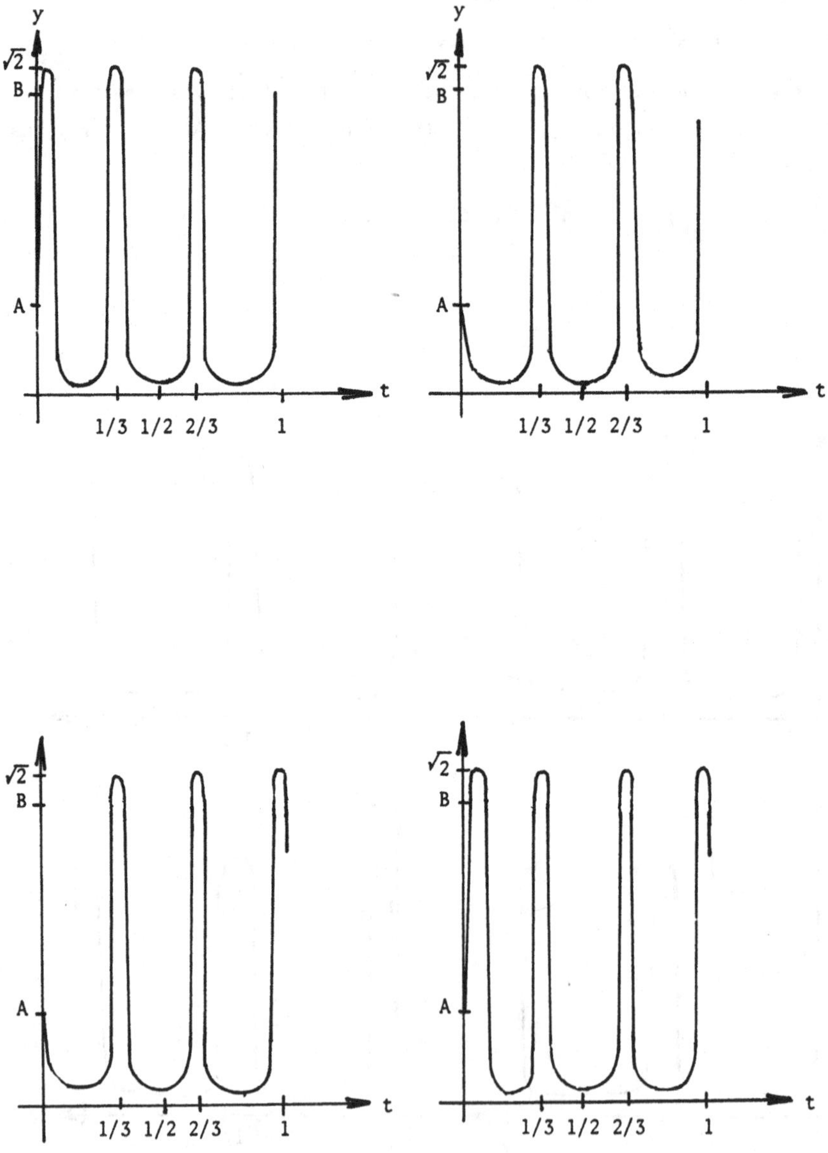

Figure 8.2
Spiked Solutions of Example 8.3 for n = 3.

with the exception that

$$\lim_{\varepsilon\to 0^+} y(t_i,\varepsilon) = \sqrt{2} \quad \text{for} \quad t_i = i/n \quad (1 \leq i \leq n-1).$$

The four solutions for the cases $n = 2$ and $n = 3$ are pictured in Figures 8.1 and 8.2.

We note finally that since $\Psi(0) = \Psi(-\sqrt{2})$, with $-\sqrt{2}$ not a maximum point of Ψ, the problem (if $|A|$, $|B| < \sqrt{2}$) also has for each integer $n \geq 2$ four solutions $y = \overline{y}(t,\varepsilon)$ as $\varepsilon \to 0^+$ satisfying

$$\lim_{\varepsilon\to 0^+} \overline{y}(t,\varepsilon) = 0 \quad \text{in} \quad [\delta, 1-\delta],$$

with the exception that

$$\lim_{\varepsilon\to 0^+} \overline{y}(t_i,\varepsilon) = -\sqrt{2}.$$

Example 8.4. We consider next the related problem

$$\varepsilon y'' = y^3 - y \equiv g(y), \quad 0 < t < 1,$$

$$y(0,\varepsilon) = A, \quad y(1,\varepsilon) = B.$$

Since $g = -h$ we see now that the reduced solutions $u_1 = 1$ and $u_3 = -1$ are stable, while $u_2 = 0$ is unstable. Let us look at the function u_1 first. The integral conditions require that (cf. Remark 3.3)

$$\int_1^{\xi} (s^3 - s)ds > 0 \quad \text{for} \quad \xi \text{ between } 1 \text{ and } A \text{ or } B,$$

and a short calculation shows this inequality holds provided $A > -1$ and $B > -1$. Consequently, if $A,B > -1$ then the problem has a solution $y = y_1(t,\varepsilon)$ as $\varepsilon \to 0^+$ such that

$$\lim_{\varepsilon\to 0^+} y_1(t,\varepsilon) = 1 \quad \text{in} \quad [\delta, 1-\delta].$$

By symmetry we note that an analogous result holds for the reduced solution $u_3 = -1$. Namely, if the boundary values A and B satisfy A, $B < 1$, then the problem has another solution $y = y_3(t,\varepsilon)$ such that

$$\lim_{\varepsilon\to 0^+} y_3(t,\varepsilon) = -1 \quad \text{in} \quad [\delta, 1-\delta].$$

In particular, we note that if $A = B = 0$ then the problem has at least *three* solutions: y_1, y_3 and $y_2 \equiv 0$.

Finally let us hasten to point out that the problem has solutions which display discontinuous interior layer (shock layer) behavior (cf. Remark 3.4). As an illustration, suppose that $A < -1$ and $B > 1$. Then O'Malley [76] has shown that the problem has a solution $y = y(t,\varepsilon)$ as $\varepsilon \to 0^+$ satisfying

$$\lim_{\varepsilon\to 0^+} y(t,\varepsilon) = \begin{cases} -1 & \text{in } [\delta, \frac{1}{2} - \delta], \\ 1 & \text{in } [\frac{1}{2} + \delta, 1-\delta], \end{cases}$$

that is, y transfers from u_3 to u_1 in a neighborhood of the point $t = 1/2$ which shrinks to zero as $\varepsilon \to 0^+$. He has also shown that when $A = B = 0$, for example, the problem has for each nonnegative integer n two solutions with limiting values 1 which switch n-times between 1 and -1. Thus, because in this example there are two stable reduced solutions separated by an unstable one, we see that there is a countably infinite number of solutions.

Application 8.1. The following boundary value problem arises as a model problem in the theory of nonpremixed combustion (cf. [97])

$$\varepsilon y'' = y^2 - t^2 \equiv h(t,y), \quad -1 < t < 1,$$

$$y(-1,\varepsilon) = y(1,\varepsilon) = 1.$$

Here ε (assumed to be very small) is a ratio of diffusive effects to the speed of reaction, and t is a distance coordinate, chosen so that $t = 0$ is the location of the flame, where the fuel and the oxidizer meet each other and react. The functions $y - t$ and $y + t$ represent the mass fractions of fuel and oxidizer, respectively.

In the limit of infinite reaction rate $\varepsilon = 0$, we obtain the reduced solutions $u_1(t) = t$ and $u_2(t) = -t$. From these we form the stable path $u(t) = |t|$ (known in combustion theory as the Burke-Schumann approximation [97]), that is, in $[-1,1]$

$$\frac{\partial h}{\partial y}(t,u(t)) \geq 0 \quad \text{and} \quad \frac{\partial^2 h}{\partial y^2} \equiv 2 > 0.$$

Theorem 3.10 then tells us that for sufficiently small $\varepsilon > 0$ the boundary value problem has a solution $y = y(t,\varepsilon)$ in $[-1,1]$ satisfying

$$y(t,\varepsilon) = |t| + O((\varepsilon^{1/3}/\sigma)(1 + \sigma|t|/\varepsilon^{1/3})^{-2}),$$

where σ is a known positive constant. Thus the thickness of the flame at $t = 0$ is of order $\varepsilon^{1/3}$.

The article of Williams [97] also discusses the same boundary value problem for more general differential equations of the form

$$\varepsilon y'' = (y^2 - t^2)^n, \quad n \geq 1, \tag{8.2}$$

and

$$\varepsilon y'' = (y+t)^n (y-t)^m, \quad m > n \geq 1. \tag{8.3}$$

Using the theory of Chapter III, the reader should have no difficulty in seeing that the thickness of the flame at $t = 0$ in the case of model (8.2) is of order $\varepsilon^{1/(2n+1)}$ and of order $\varepsilon^{1/(m+n+1)}$ in the case of model (8.3).

We turn now to a consideration of some related Robin problems.

Example 8.5. The first Robin problem is

$$\varepsilon y'' = (y - u(t))^{2q+1}, \quad -1 < t < 1,$$

$$y(-1,\varepsilon) - y'(-1,\varepsilon) = A, \quad y(1,\varepsilon) + y'(1,\varepsilon) = B,$$

where q is a nonnegative integer. If the function u is, say, twice continuously differentiable in $[-1,1]$, then Theorem 3.7 tells us that there is a solution $y = y(t,\varepsilon)$ as $\varepsilon \to 0^+$ such that

$$\lim_{\varepsilon \to 0^+} y(t,\varepsilon) = u(t) \quad \text{in} \quad [-1,1]. \tag{8.4}$$

On the other hand, if u is a function such as $|t|$, then the same relation (8.4) still holds; however, there is an angular layer at $t = 0$ whose thickness is of order $\varepsilon^{1/(2q+2)}$ (cf. Theorem 3.9).

In this example, we do not need any restriction on the boundary values, in order that the uniform limit (8.4) holds. The next example shows that restrictions are sometimes needed on the boundary values.

Example 8.6. Consider the Robin problem

$$\varepsilon y'' = (y - t^2)^2, \quad 0 < t < 1,$$

$$y(0,\varepsilon) - y'(0,\varepsilon) = A, \quad y(1,\varepsilon) + y'(1,\varepsilon) = B.$$

Owing to the quadratic nature of the righthand side, we require that the reduced solution $u(t) = t^2$ satisfy $u(0) - u'(0) \leq A$ and $u(1) + u'(1) \leq B$, that is, the boundary values must satisfy $A \geq 0$ and $B \geq 3$. For such values of A and B, Theorem 3.8 states that for sufficiently small $\varepsilon > 0$, the problem has a solution satisfying

$$\lim_{\varepsilon \to 0^+} y(t,\varepsilon) = t^2 \quad \text{in} \quad [0,1].$$

Example 8.7. We consider next some Robin analogs of Examples 8.3 and 8.4, namely

$$\varepsilon y'' = y - y^3 \tag{8.5}$$

and

$$\varepsilon y'' = y^3 - y \tag{8.6}$$

for t in $(0,1)$, along with the boundary conditions $y(0,\varepsilon) - y'(0,\varepsilon) = A$, $y(1,\varepsilon) + y'(1,\varepsilon) = B$. In the case of (8.5) we know that $u = 0$ is the only stable reduced solution, and so Theorem 3.7 tells us that for sufficiently small $\varepsilon > 0$, the Robin problem (8.5) has a solution satisfying

$$\lim_{\varepsilon \to 0^+} y(t,\varepsilon) = 0 \quad \text{in} \quad [0,1].$$

Similarly, since both $u_1 = 1$ and $u_3 = -1$ are stable reduced solutions of (8.6), this same theorem states that for sufficiently small $\varepsilon > 0$, the Robin problem (8.6) has two solutions $y_1(t,\varepsilon)$ and $y_2(t,\varepsilon)$ satisfying in $[0,1]$

$$\lim_{\varepsilon \to 0^+} y_1(t,\varepsilon) = 1$$

and

$$\lim_{\varepsilon \to 0^+} y_2(t,\varepsilon) = -1.$$

However we are unable to say at the present time whether there are other solutions of (8.6) switching between 1 and -1 such as those found in Example 8.4 using O'Malley's results. Such an occurrence would be very interesting mathematically, and it would also have important consequences for several problems in catalytic reaction theory, the subject of the next application.

Application 8.2. Our final example in this section is an application of our results to a class of boundary value problems arising in catalytic reaction theory (see, for example, [2; Chapter 3]). The simplified physical problem involves an isothermal reaction $A \to B$ which is catalyzed in a pellet of length two. An equation that describes the mass balance between diffusion and reaction inside of the pellet is then

$$y'' = \Phi^2 R(y), \quad 0 < t < 1,$$

where y is the normalized concentration of the reactant A, t is the (dimensionless) distance from the center of the pellet ($t = 0$) to the mouth ($t = 1$), Φ is the Thiele modulus which measures the effect of diffusion as opposed to reaction, and $R(y)$ is the reaction rate term. In particular, Φ^2 is proportional to k/D, where D is the diffusion coefficient and k the reaction rate constant. Let us assume that $R(y) = y^n$, for a nonnegative integer n, that is, the reaction is nth order, and that the catalytic reaction is diffusion-limited, that is, $\Phi^2 \gg 1$. At the line of symmetry ($t = 0$) there is no flux, that is, $y'(0,\varepsilon) = 0$, while at the mouth of the pellet the correct boundary condition is $y(1,\varepsilon) + \Sigma y'(1,\varepsilon) = 1$, for a nonnegative constant Σ whose reciprocal is called the Sherwood number. The Sherwood number measures the ability of the reactant to reach the pellet from the bulk flow. Incorporating all of this into the model, we arrive finally at the problem

$$\varepsilon y'' = y^n, \quad 0 < t < 1,$$

$$-y'(0,\varepsilon) = 0, \quad y(1,\varepsilon) = \quad y'(1,\varepsilon) = 1,$$

where $\varepsilon = \Phi^{-2} \ll 1$.

Let us consider first the case when $\Sigma = 0$, that is, the pellet is uniformly accessible to the bulk flow. Theorems 3.4 and 3.5 then tell us that the boundary value problem has a nonnegative solution satisfying

$$\lim_{\varepsilon \to 0^+} y(t,\varepsilon) = 0 \quad \text{in} \quad [0,1-\delta]. \tag{8.7}$$

Qualitatively this means that owing to the large rate of reaction (or equivalently, small rate of diffusion), most of the reaction is confined to the pore mouth. The concentration of reactant falls off to zero exponentially (algebraically) for $n = 1$ ($n \geq 2$), as we proceed into the pellet.

Suppose we study now the more realistic case $\Sigma > 0$, in which there is some resistance to the transfer of reactant from the bulk to the pellet. If the resistance is small, that is, if $\Sigma = O(\varepsilon)$, then we again have the situation described by (8.7). However if Σ is asymptotically larger than ε, say $\Sigma = O(1)$, then Theorems 3.7 and 3.8 state that for sufficiently small $\varepsilon > 0$ the boundary value problem has a nonnegative solution satisfying

$$\lim_{\varepsilon \to 0^+} y(t,\varepsilon) = 0 \quad \text{in} \quad [0,1].$$

Qualitatively this means that in the presence of resistance to mass transfer, the concentration of reactant in the reaction zone near the pore mouth $t = 1$ vanishes as $\varepsilon \to 0^+$. More precisely, we have $y(1,\varepsilon) = O(\varepsilon)$ if $n = 1$, and $y(1,\varepsilon) = O(\varepsilon^{1/(n+1)})$ if $n \geq 2$, by virtue of the estimates in these theorems.

§8.2. Examples of Quasilinear Problems and Applications

Example 8.8. We begin with the simplest quasilinear problem

$$\varepsilon y'' = f(y)y', \quad a < t < b,$$

$$y(a,\varepsilon) = y_-, \quad y(b,\varepsilon) = y_+, \quad y_- < y_+,$$

where f is continuous for $y_- \leq y \leq y_+$. Let us assume that the reduced solutions $u_L = y_-$ and $u_R = y_+$ are stable, in the sense that $f(y_+) < 0 < f(y_-)$, and let us define the functional $J[y] \equiv \int_{y_-}^{y} f(s)ds = F(y) - F(y_-)$, where F is an antiderivative of f. Then the theory of Chapter IV (refer also to Remark 4.3) asserts that if $J[y_+] < 0$, that is, if $F(y_+) < F(y_-)$, the problem has a solution $y = y(t,\varepsilon)$ as $\varepsilon \to 0^+$ satisfying

$$\lim_{\varepsilon\to 0^+} y(t,\varepsilon) = y_+ \quad \text{in} \quad [a+\delta,b].$$

On the other hand, if $F(y_+) > F(y_-)$, then $J[y_+] > 0$, and the theory tells us that the solution of the problem satisfies

$$\lim_{\varepsilon\to 0^+} y(t,\varepsilon) = y_- \quad \text{in} \quad [a,b-\delta].$$

There remains the case when $F(y_+) = F(y_-)$. The existence of boundary layers is now precluded, and so we look for another type of limiting behavior as $\varepsilon \to 0^+$. (Note that the boundary value problem has a solution for all $\varepsilon > 0$ by virtue of Theorem 2.1, which is unique, since $\alpha(t,\varepsilon) \equiv y_-$ and $\beta(t,\varepsilon) \equiv y_+$ are lower and upper solutions, respectively.) Because the reduced solutions are constants, the only type of behavior possible is that involving a shock layer connecting the states y_- and y_+ at some point t_0 in (a,b). In order to locate this transition point, let us begin by noting that the solution satisfies

$$y'(t,\varepsilon) = \text{const.} \exp[F(y(t,\varepsilon))], \tag{8.8}$$

and so $y'(t,\varepsilon) > 0$ since by assumption $y_- < y_+$. Thus we can rewrite the original differential equation in the form

$$\varepsilon y''/y' = \varepsilon(\ln y')' = f(y),$$

which in turn allows us to write the two equations

$$\varepsilon \ln y'(t_0,\varepsilon) - \varepsilon \ln y'(a,\varepsilon) = \int_a^{t_0} f(y(s,\varepsilon))ds \tag{8.9}$$

and

$$\varepsilon \ln y'(b,\varepsilon) - \varepsilon \ln y'(t_0,\varepsilon) = \int_{t_0}^{b} f(y(s,\varepsilon))ds. \tag{8.10}$$

But $y'(a,\varepsilon) = y'(b,\varepsilon)$ by virtue of (8.8), since $F(y_-) = F(y_+)$, and so adding (8.9) and (8.10) gives

$$0 = \int_a^{t_0} f(y(s,\varepsilon))ds + \int_{t_0}^{b} f(y(s,\varepsilon))ds. \tag{8.11}$$

Finally, since

$$\lim_{\varepsilon\to 0^+} y(t,\varepsilon) = \begin{cases} y_-, & a \le t \le t_0-\delta, \\ y_+, & t_0+\delta \le t \le b, \end{cases}$$

(that is, there is a shock layer at t_0), if we take the limit as $\varepsilon \to 0^+$ of both sides of (8.11), we obtain from the Dominated Convergence Theorem the limiting relation

$$0 = \int_a^{t_0} f(y_-)ds + \int_{t_0}^{b} f(y_+)ds = f(y_-)(t_0 - a) + f(y_+)(b - t_0).$$

It follows that the shock layer is located at

$$t_0 = [f(y_-)a - f(y_+)b]/[f(y_-) - f(y_+)].$$

For example, if $f(y) = -y$, then in order for there to be a shock layer connecting y_- to y_+, we must have $y_+ = -y_- > 0$; the layer is located at $t_0 = (a+b)/2$.

The relation $F(y_-) = F(y_+)$, known as the Rankine-Hugoniot shock condition (cf. [17; Chapter 3], [55; Chapter 4]), arises in modelling compressible flows and chemically reacting flows, as shown in the next application.

Application 8.3. This problem concerns the description of the one-dimensional, steady-state flow pattern arising from the injection of a gas at supersonic velocity into a duct of uniform or diverging cross-sectional area when a back pressure is applied. Complications such as the effect of viscous stresses on the duct walls is neglected, and the gas is assumed to be perfect and polytropic. The time-independent laws of conservation of mass, momentum and energy can be expressed in the following dimensionless form by referring all quantities to appropriate lengths, physical constants and upstream conditions (cf. [18]):

$$d/dx(\rho yA) = 0, \tag{8.12}$$

$$y\frac{dy}{dx} + (\gamma\rho)^{-1}\frac{d}{dx}(\rho T) = \mu\rho^{-1}\frac{d^2y}{dx^2}, \tag{8.13}$$

and

$$y\frac{dT}{dx} + (\gamma-1)T[\frac{dy}{dx} + y\frac{d}{dx}(\ln A)] - \gamma(\gamma-1)\mu\rho^{-1}(\frac{dy}{dx})^2$$
$$= \mu\gamma\rho^{-1}P_r^{-1}\frac{d^2T}{dx^2}. \tag{8.14}$$

Here x is the dimensionless distance measured from the entrance of the duct, y is the dimensionless velocity of the gas relative to the velocity of sound, ρ is the density, γ is the adiabatic index with a value between 1 and 5/3, T is the dimensionless temperature, μ is the coefficient of viscosity, and P_r is the Prandtl number, taken equal to 3/4. Finally $A = A(x)$ $(A(0) = 1)$ is the dimensionless cross-sectional area of the duct relative to the area of the duct entrance. Following Crocco [18] we have omitted terms of the form $\mu dA/dx$ and $\mu P_r^{-1}\,dA/dx$ in (8.13) and (8.14), respectively. By first neglecting the second-order stress terms in these equations we obtain easily two equations for isentropic flow (cf. [18])

$$A\bar{y}[1 - \frac{(\gamma-1)}{2}\bar{y}^2]^{1/(\gamma-1)} = \text{const.}$$

and

$$\bar{T} = 1 - \frac{(\gamma-1)}{2}\bar{y}^2 = (\bar{\rho})^{\gamma-1}.$$

Upon substituting the expression $T(x) = 1 - \frac{(\gamma-1)}{2}y^2(x)$ into (8.13) and using (8.12) we obtain, after a straightforward calculation, an equation for the velocity y of the form

$$\mu\gamma(\rho_0c_0)^{-1}Ay\frac{d^2y}{dx^2} = [\frac{(\gamma+1)}{2}y - y^{-1}]\frac{dy}{dx} - \frac{d}{dx}[\ln A(1 - \frac{(\gamma-1)}{2}y^2)].$$

The quantities ρ_0, c_0 are respectively an upstream reference density and the upstream velocity of sound. Let us now assume that the dimensionless term $\mu\gamma(\rho_0c_0)^{-1} = \varepsilon$ is small, that is, the coefficient of viscosity μ is small for fixed values of γ, ρ_0 and c_0. Then we have the singularly perturbed quasilinear problem in $(0,1)$

$$\varepsilon Ay\frac{d^2y}{dx^2} = [\frac{(\gamma+1)}{2}y - y^{-1}]\frac{dy}{dx} - \frac{d}{dx}[\ln A(1 - \frac{(\gamma-1)}{2}y^2)],$$
$$y(0,\varepsilon) = y_-, \quad y(1,\varepsilon) = y_+, \quad y_- > y_+ > 0. \tag{SL}$$

(This is basically the same boundary value problem used by Pearson [77] in his numerical experiments with $A(x) = 1 + x^2$.) The original physical problem can be restated now in terms of (SL) as follows: given a supersonic velocity y_- at the entrance of the duct ($x = 0$), determine what subsonic velocity y_+ at the end of the duct ($x = 1$) produces a supersonic-subsonic transition in the interior of the duct and also the location of this transition.

Let us consider first the case of a uniform duct, that is, $A(x) \equiv 1$. The problem (SL) reduces to the simple form

$$\varepsilon\frac{d^2y}{dx^2} = [\frac{(\gamma+1)}{2} - y^{-2}]\frac{dy}{dx}, \quad 0 < x < 1,$$
$$y(0,\varepsilon) = y_-, \quad y(1,\varepsilon) = y_+. \tag{SL_0}$$

In order to find conditions under which (SL_0) has a shock layer solution, we have only to apply the results of Example 8.8. For $y > 0$ the function $f(y) = \frac{(\gamma+1)}{2} - y^{-2}$ is continuous and vanishes only at $y_c = (2/\gamma+1)^{1/2}$, which can be regarded as a dimensionless critical velocity [17; Chapter 3]. Consequently, $f(y) > 0$ for $y > y_c$ and $f(y) < 0$ for $0 < y < y_c$. The Rankine-Hugoniot shock condition is

$$\frac{(\gamma+1)}{2}y_- + y_-^{-1} = \frac{(\gamma+1)}{2}y_+ + y_+^{-1},$$

that is,

$$y_+y_- = y_c^2 = \frac{2}{(\gamma+1)},$$

which is known as Prandtl's relation (cf. [17; Chapter 3]). Thus, given an initial supersonic velocity $y_- > y_c$, there is a unique subsonic velocity $y_+ = \frac{2}{(\gamma+1)}y_-^{-1}$ such that (SL_0) has a solution satisfying

$$\lim_{\varepsilon\to 0^+} y(x,\varepsilon) = \begin{cases} y_-, & 0 \le x \le x_0-\delta, \\ y_+, & x_0+\delta \le x \le 1. \end{cases}$$

Here

$$x_0 = \frac{f(y_+)}{f(y_+)-f(y_-)} = \frac{y_-}{y_- + y_+}$$

is the location of the shock layer representing a supersonic-subsonic transition. This formula for x_0 allows us to conclude that if the supersonic inlet velocity is very large, then the major portion of the flow is supersonic since x_0 is close to unity. The shock layer sits close to the end of the duct.

We turn finally to a consideration of (SL) when the cross-sectional area of the duct increases in the downstream direction, that is, $dA/dx > 0$ for $0 < x \le 1$. Setting $\varepsilon = 0$ we first obtain solutions of the reduced equation $(\frac{(\gamma+1)}{2} u - u^{-1})\frac{du}{dx} = \frac{d}{dx}[\ln A(1 - \frac{(\gamma-1)}{2} u^2)]$ satisfying $u_1(0) = y_-$ and $u_2(1) = y_+$ implicitly as

$$u_1(x)(1 - \frac{(\gamma-1)}{2} u_1^2(x))^{1/(\gamma-1)} = y_-(1 - \frac{(\gamma-1)}{2} y_-^2)^{1/(\gamma-1)}/A(x)$$

and

$$u_2(x)(1 - \frac{(\gamma-1)}{2} u_2^2(x))^{1/(\gamma-1)} = A(1)y_+(1 - \frac{(\gamma-1)}{2} y_+^2)^{1/(\gamma-1)}/A(x),$$

respectively. (Recall that $A(0) = 1$.) Not unexpectedly, these are the isentropic relations obtained at the beginning of our discussion. Since $f(y) = \frac{(\gamma+1)}{2} - y^{-2}$ is positive (or negative) for $y > y_c = (\frac{2}{(\gamma+1)})^{1/2}$ (or $0 < y < y_c$), we must require that $u_1(x) > y_c$ in $[0,x_1]$ and $0 < u_2(x) < y_c$ in $[x_2,1]$ with $x_1 > x_2$. Then our theory applies provided that the Rankine-Hugoniot equation $F(u_1(x_0)) = F(u_2(x_0))$, for $F(y) = \frac{(\gamma+1)}{2} y + y^{-1}$, has a solution x_0 in the interval (x_2,x_1). A short calculation shows that this condition is equivalent to $u_1(x_0)u_2(x_0) = y_c^2 = \frac{2}{(\gamma+1)}$, that is, Prandtl's relation must hold at x_0. We conclude that under these assumptions, the problem (SL) has a solution satisfying

$$\lim_{\varepsilon\to 0^+} y(x,\varepsilon) = \begin{cases} u_1(x), & 0 \le x \le x_0-\delta, \\ u_2(x), & x_0+\delta \le x \le 1. \end{cases}$$

The solution describes a supersonic-subsonic transition at x_0 between the nonconstant states $u_1(x)$ and $u_2(x)$.

In order to illustrate this result, let us consider the case $A(x) = 1 + x^2$, $y_- = 0.9129$, $y_+ = 0.375$ and $\gamma = 7/5$. The problem (SL), with this data and with the term $-\varepsilon y \frac{dy}{dx}$ added to the righthand side, was treated numerically by Pearson [77] whose results afford a means of comparison with ours. For $\gamma = 7/5$ the critical velocity $y_c = (\frac{2}{(\gamma+1)})^{1/2}$ is slightly less than y_-, and so $y_- > y_c > y_+$. With this choice of y_+ and with $A(x) = 1 + x^2$ the reduced solution u_2 assumes the critical value y_c at a point x_2 in $(0,1/2)$, that is, $f(u_2(x_2)) = 0$. As regards u_1, it is easy to see that $u_1(x) > y_c$; whence, $f(u_1(x)) > 0$ in $[0,1]$. Finally, one can show that the Rankine-Hugoniot equation $F(u_1(x_0)) = F(u_2(x_0))$ has a unique solution x_0 in $(x_2,1)$ which is approximately 0.6. This compares well with the value $x_0 = 0.634$ obtained numerically by Pearson for the slightly modified version of (SL) with ε of order 10^{-8}.

Example 8.9. The reasoning employed in Example 8.8 extends to more general problems of the form

$$\varepsilon y'' = f(y)y' + g(y), \quad a < t < b,$$

$$y(a,\varepsilon) = y_-, \quad y(b,\varepsilon) = y_+,$$

provided that the boundary values are solutions of the reduced equation, that is, $g(y_-) = g(y_+) = 0$. For instance, we know from Example 8.8 that for $a = y_- = -1$, $b = y_+ = 1$, $f(y) = -y$ and $g \equiv 0$, this problem has a solution connecting y_- to y_+ across a shock layer at $t = 0$, as $\varepsilon \to 0^+$. However, if $g(y) = 1 - y^2$, then this same result also holds. Asymptotically the g-term has no effect on the behavior of solutions, provided that the f-term has the properties given above.

Example 8.10. Consider now the problem

$$\varepsilon y'' = -y^n y' + y, \quad 0 < t < 1,$$

$$y(0,\varepsilon) = A, \quad y(1,\varepsilon) = B,$$

where n is a positive real number. When $n = 1$, it is the classic Lagerstrom-Cole model problem about which much has been written (cf. [55; Chapter 2], [20], [39]). Let us look first for a solution with a boundary layer at $t = 0$. If $B^n > n$ then the function $u_R(t) = (n(t-1) + B^n)^{1/n}$ is a strongly stable solution of the corresponding righthand reduced problem. Suppose now that n is a natural number. There are two cases to consider. First, if n is even, then $-y^n$, the coefficient of y', is nonpositive throughout the layer, and so Theorem 4.1 implies that the problem has a solution satisfying

$$\lim_{\varepsilon\to 0^+} y(t,\varepsilon) = u_R(t) \quad \text{in } [\delta,1], \quad \delta > 0, \tag{8.15}$$

for all values of A. Second, if n is odd, then the relative position of A and $u_R(0)$ is important. If $A \geq u_R(0) = (B^n - n)^{1/n}$ or if $0 \leq A < u_R(0)$, then the coefficient of y' is again nonpositive in the layer, and so the limit (8.15) also holds for such values of A. However, if $A < 0$, then we must apply the integral condition of Coddington and Levinson [14] (cf. Remark 4.3) which allows values of A for which

$$\int_{\xi}^{u_R(0)} -s^n ds < 0, \quad A \leq \xi < u_R(0).$$

An easy calculation reveals that this inequality holds if $|A| < (B^n-n)^{1/n}$. Thus for these values of A, the limiting relation (8.15) is also valid. In summary, if n is even there is a boundary layer at $t = 0$ for all values of A, while if n is odd there is a layer only if $A > -(B^n-n)^{1/n}$. A moment's reflection shows that these conclusions hold when $B^n = n$ (cf. Theorem 4.2).

Suppose next that $0 < B^n < n$. Then the function $u_R(t) = (n(t-1) + B^n)^{1/n}$ vanishes at $t_R = 1 - B^n/n$ in $(0,1)$. If $0 < n < 1$ then $u_R(t_R) = u_R'(t_R) = 0$, that is, u_R intersects the zero reduced solution smoothly. On the other hand, if $n > 1$ then $u_R'(t_R^+) = \infty$, and it is not clear how to proceed. However, for the intermediate case $n = 1$, $u_R'(t_R) = 1$, and we can say a few words about this case. The zero reduced solution is clearly (I_0)-stable (cf. Definition 4.4), and so the reduced path

$$u_0(t) = \begin{cases} 0, & 0 \leq t \leq t_R = 1-B, \\ t+B-1, & t_R \leq t \leq 1, \end{cases}$$

is weakly stable (cf. Definition 4.5). If $A > 0$, then Theorem 4.12 tells us that for sufficiently small $\varepsilon > 0$, the problem has a solution satisfying

$$\lim_{\varepsilon\to 0^+} y(t,\varepsilon) = u_0(t) \quad \text{in} \quad [\delta,1].$$

(Of course, if $A = 0$ then this limit is assumed at $t = 0$ as well.) Next, if $-1 < A < 0$ and $-A < 1-B$ then the weakly stable lefthand reduced solution $u_L(t) = A + t$ intersects the zero solution at $t_L = -A < t_R$, and we have a situation described by Theorem 4.14. The problem has a solution satisfying

$$\lim_{\varepsilon\to 0^+} y(t,\varepsilon) = \begin{cases} t+A, & 0 \leq t \leq t_L, \\ 0\;, & t_L \leq t \leq t_R, \\ t+B-1, & t_R \leq t \leq 1. \end{cases}$$

When $-A = 1-B$ note that $y(t,\varepsilon) = t+A = t+B-1$ is the exact solution! Finally, for $-1 < A < 0$ and $-A > 1-B$, there are no angular crossings since $t_L > t_R$. In this case, one can show (cf. [20], [39]) that the solution as $\varepsilon \to 0^+$ connects u_L and u_R across a shock layer at $t_0 = 1/2(1-B-A)$ in (t_R,t_L). Finally, for $A < -1$ $(A = -1)$ the function u_L is strongly (weakly) stable in $[0,1]$, and so there are solutions with boundary layers at $t = 1$ or with shock layers at t_0 joining u_L and u_R, depending on the relative sizes of A and B. See [20] or [39] for all of the details.

The last phenomenon we discuss is the existence of boundary layers relative to the (I_0)-stable zero function, when n is a natural number. If n is even and $B = 0$, then Theorem 4.5 tells us that the problem has a solution satisfying

$$\lim_{\varepsilon\to 0^+} y(t,\varepsilon) = 0 \quad \text{in} \quad [\delta,1]$$

for all values of A, with $\delta = 0$ if $A = 0$. However, if n is odd, then by Theorem 4.5 we have for all values of $A \geq 0$ and $B \leq 0$, a solution satisfying

$$\lim_{\varepsilon\to 0^+} y(t,\varepsilon) = 0 \quad \text{in} \quad [\delta,1-\delta].$$

For such values of A, B and/or n, either u_L or u_R does not exist or is unstable.

Application 8.4. For our final application of this section we consider a catalytic reaction in a one-dimensional fixed-bed reactor packed with catalyst in the presence of axial diffusion. The boundary value problem for the dimensionless concentration y is then (cf. [80; Chapter 4])

$$\varepsilon y'' = y' + g(y), \quad 0 < x < 1,$$

$$y(0,\varepsilon) - p_1 y'(0,\varepsilon) = A, \quad y(1,\varepsilon) = p_2 y'(1,\varepsilon) = B,$$

where x is the dimensionless axial coordinate, ε is the reciprocal of the Peclet number (the diffusional analog of the Reynolds number), and g is the reaction rate term, of the form $g(y) = y^n$ for an n-th order reaction. The positive terms $p_1 = p_1(\varepsilon)$ and $p_2 = p_2(\varepsilon)$ are mass transport coefficients of the type considered in Application 8.2. If the axial diffusion is weak then ε is small, and the Robin problem is singularly perturbed.

We consider only two cases. First, if $p_2(\varepsilon) = O(\varepsilon)$ then Corollary 4.2 tells us that the problem has a solution with a boundary layer at $x = 1$, that is,

$$\lim_{\varepsilon\to 0^+} y(x,\varepsilon) = u_L(x) \quad \text{in} \quad [0,1-\delta],$$

where u_L, the solution of the reduced problem $u' + g(u) = 0$, $u(0) - p_1u'(0) = A$, is assumed to exist in $[0,1]$. Second, if $p_2 = O(1)$ or larger, then Corollary 4.3 tells us that there is a solution which satisfies this limiting relation in all of $[0,1]$. Here we are taking $g(y) = y^n$, for n a natural number.

Depending on the nature of the function g, O'Malley [75; Chapter 7] has shown that this problem may have more than one solution. Multiplicity of solutions, which are steady states of the associated time-dependent partial differential equation, is often an important consideration in reactor design, since it implies that the reactor may be capable of operating in more than one stable regime.

Example 8.11. We conclude this section with a problem about which we know very little, namely

$$\varepsilon y'' = ty' + y^3 - y, \quad -1 < t < 1,$$

$$y(-1,\varepsilon) - y'(-1,\varepsilon) = -1, \quad y(1,\varepsilon) + y'(1,\varepsilon) = 1.$$

Clearly $u_L \equiv -1$ and $u_R \equiv 1$ are solutions of the corresponding left- and righthand reduced problems, respectively. Since these functions are (I_0)-stable, we can apply Theorem 4.10 to deduce the existence of two solutions y_1, y_2 of the problem as $\varepsilon \to 0^+$ satisfying in $[-1,1]$

$$\lim_{\varepsilon\to 0^+} y_1(t,\varepsilon) = -1$$

and

$$\lim_{\varepsilon\to 0^+} y_2(t,\varepsilon) = 1.$$

There are however other solutions of the reduced equation -- the zero function which is unstable and the one-parameter family of functions $u_c(t) = ct/(1 + c^2t^2)^{1/2}$, where c is a constant, as shown in Figure 8.3. The functions u_c are all capable of supporting boundary layers at the endpoints, since the coefficient of y' is negative (positive) near $t = -1$ $(t = 1)$, but they lose stability upon passing through the origin. We strongly suspect, though, that the u_c's do participate in describing the asymptotic behavior of other solutions of the problem. It might be instructive to collect some numerical data in this direction.

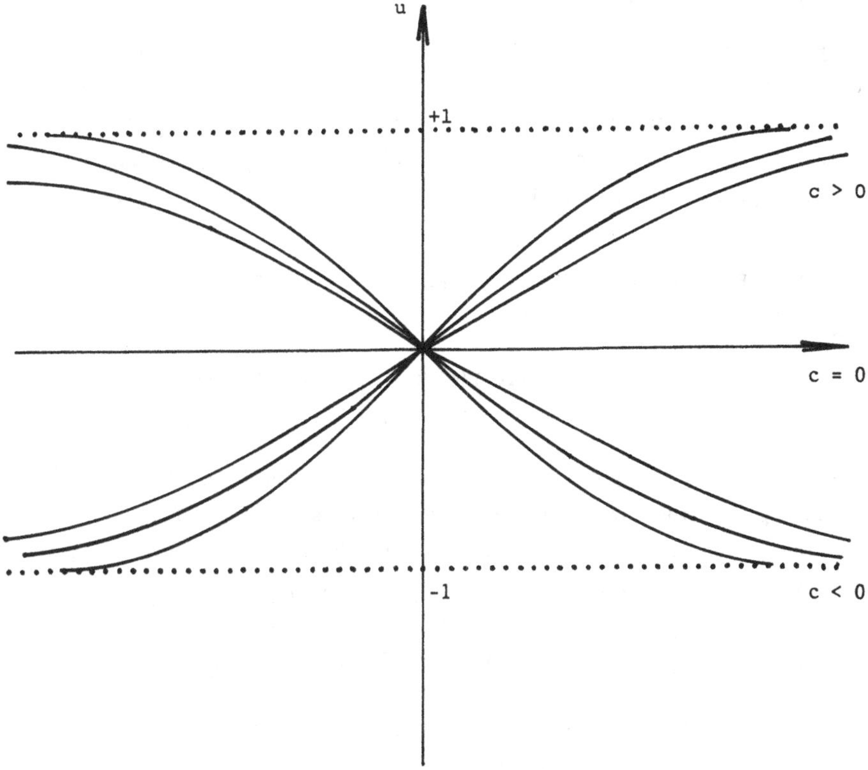

Figure 8.3
Solutions of the Reduced Equation of Example 8.11

§8.3. Examples of Quadratic Problems and Applications

Example 8.12. Consider the problem

$$\varepsilon y'' = -y'^2 - yy', \quad 0 < t < 1,$$

$$y(0,\varepsilon) = A, \quad y(1,\varepsilon) = B,$$

treated by Dorr, Parter and Shampine [20; §5] in the case $A < B$. Our first step is to examine the reduced equation $0 = u'(u'+u)$, and we find that $u_L(t) \equiv A$, $\bar{u}_L(t) = Ae^{-t}$, $u_R(t) \equiv B$ and $\bar{u}_R(t) = Be^{1-t}$ are four solutions of the corresponding reduced problems. Note that this reduced equation has no singular solutions. Setting $f(t,y,y') = -y'^2 - yy'$, we see that $f_{y'} = -2y' - y$, and so $f_{y'}(t,u_L,u_L') = f_{y'}[u_L] = -A$, $f_{y'}[\bar{u}_L] = Ae^{-t}$, $f_{y'}[u_R] = -B$ and $f_{y'}[\bar{u}_R] = Be^{1-t}$. Thus the stability of these reduced solutions depends critically on the values of A and B. We distinguish the following cases.

Case 1. $A,B > 0$. Clearly for such A, B, $\bar{u}_L(t) = Ae^{-t}$ and $u_R(t) \equiv B$ are the stable reduced solutions. We then check the sign restrictions necessary for boundary layer behavior, that is, $\bar{u}_L(1) = Ae^{-1} \geq B$ and $u_R(0) = B \geq A$. For these values of A, B, it follows from Theorem 5.1 that the problem has a solution $y = y(t,\varepsilon)$ as $\varepsilon \to 0^+$ such that in $[0,1]$

$$B - (B-A)\exp[-Bt/\varepsilon] \leq y(t,\varepsilon) \leq B, \quad \text{if } B \geq A > 0,$$

and

$$Ae^{-t} - (Ae^{-1}-B)\exp[-Ae^{-1}(1-t)/\varepsilon] \leq y(t,\varepsilon) \leq Ae^{-t} + \varepsilon\gamma,$$
$$\text{if } Ae^{-1} \geq B > 0,$$

where γ is a known positive constant. We next check for the occurrence of an angular crossing involving these reduced solutions. That is, we consider values of $A, B > 0$ for which the above boundary layer inequalities fail simultaneously; namely, $Ae^{-1} < B$ and $B < A$. It is easy to see that $\bar{u}_L(t_0) = u_R(t_0) = B$ for $t_0 = \ln(A/B)$ in $(0,1)$, with $\bar{u}_L'(t_0) = -B < u_R'(t_0) = 0$, and since the inequality $f(t_0,\bar{u}_L(t_0),\omega) = -\omega(\omega+B) > 0$ for ω in $(-B,0)$, the theorem of Haber and Levinson [27] (of. Corollary 5.20) applies. We conclude that if $0 < Ae^{-1} < B < A$, the problem has a solution satisfying

$$\lim_{\varepsilon\to 0^+} y(t,\varepsilon) = \begin{cases} Ae^{-t}, & 0 \leq t \leq t_0, \\ B, & t_0 \leq t \leq 1. \end{cases}$$

Case 2. $A > 0$, $B \leq 0$. In this case, the functions $\bar{u}_L(t) = Ae^{-t}$ and $\bar{u}_R(t) = Be^{1-t}$ are the stable reduced solutions. Since $B \leq 0 < A$ we see that only the boundary layer inequality $\bar{u}_L(1) > B$ obtains, and so by Theorem 5.1 the problem has a solution satisfying

$$\lim_{\varepsilon\to 0^+} y(t,\varepsilon) = Ae^{-t} \quad \text{in } [0,1-\delta].$$

In this case there are no interior crossings.

Case 3. $A \leq 0$, $B > 0$. This is the reflection of Case 2 and so the solution satisfies

$$\lim_{\varepsilon\to 0^+} y(t,\varepsilon) = B \quad \text{in } [\delta,1].$$

Case 4. $A, B < 0$. This is also a reflection, of Case 1, in which the reduced solutions $u_L(t) \equiv A$ and $\bar{u}_R(t) = Be^{1-t}$ are stable. We see easily that the problem has solutions with the following asymptotic behavior:

$$\lim_{\varepsilon\to 0^+} y(t,\varepsilon) = A \quad \text{in} \quad [0,1-\delta], \quad \text{if } A \geq B,$$

$$\lim_{\varepsilon\to 0^+} y(t,\varepsilon) = Be^{1-t} \quad \text{in} \quad [\delta,1], \quad \text{if} \quad Be \geq A,$$

and finally

$$\lim_{\varepsilon\to 0^+} y(t,\varepsilon) = \begin{cases} A, & 0 \leq t \leq \bar{t}_0, \\ Be^{1-t}, & \bar{t}_0 \leq t \leq 1, \end{cases} \quad \text{if} \quad Be < A < B,$$

for $\bar{t}_0 = \ln(Be/A)$ in $(0,1)$.

Two cases remain: (i) $A < 0, B = 0$ and (ii) $A = 0, B < 0$. In the first case, $u_L \equiv A$ is strongly stable; however, since $B = 0$, $u_R = \bar{u}_R \equiv 0$ and $f_{y'}[0] = f_y[0] \equiv 0$. But $u_L(1) = A < B$ and u_L does not intersect u_R since $u_L < 0$. Thus u_L does not determine the asymptotic behavior in this case. Let us examine $u_R \equiv 0$ more closely. Clearly, $\beta \equiv 0$ is an upper solution of the problem, and a short calculation shows that

$$\alpha(t,\varepsilon) = \begin{cases} (A^{-1} - t/\varepsilon^{1/2})^{-1} - \gamma t\varepsilon^{1/2}, & 0 \leq t \leq 1-\tau(\varepsilon), \\ (A^{-1} - (1-\tau(\varepsilon))/\varepsilon^{1/2})^{-1} - \gamma(1-\tau(\varepsilon))\varepsilon^{1/2}, & 1-\tau(\varepsilon) \leq t \leq 1, \end{cases}$$

is a lower solution for $\tau(\varepsilon) > 0$ of order $\varepsilon^{1/2}$, provided that $\gamma > 0$ is appropriately chosen. We conclude from Theorem 2.2 that if $A < 0$ and $B = 0$ the problem has, for sufficiently small ε, a solution satisfying $\alpha(t,\varepsilon) \leq y(t,\varepsilon) \leq 0$ in $[0,1]$, that is,

$$\lim_{\varepsilon\to 0^+} y(t,\varepsilon) = 0 \quad \text{in} \quad [\delta,1].$$

Arguing in a similar manner, we see that in case (ii) the problem has a solution satisfying

$$\lim_{\varepsilon\to 0^+} y(t,\varepsilon) = 0 \quad \text{in} \quad [0,1-\delta].$$

These two cases deserve to be given special care, since appropriate partial derivatives are neither strictly positive nor negative.

Example 8.13. The problem

$$\varepsilon y'' = -y'^2 + y, \quad 0 < t < 1,$$

$$y(0,\varepsilon) = A, \quad y(1,\varepsilon) = B,$$

illustrates how the existence of a singular reduced solution affects the asymptotic nature of solutions of a quadratic problem for certain values of the boundary data. Its reduced equation $u'^2 = u$ is a Clairaut equation which has the regular solutions

$$u_L(t) = 1/4(2A^{1/2} - t)^2, \quad u_R(t) = 1/4(t + 2B^{1/2} - 1)^2$$

and

$$\bar{u}_L(t) = 1/4(2A^{1/2} + t)^2, \quad \bar{u}_R(t) = 1/4(2B^{1/2} + 1 - t)^2,$$

defined for $A, B \geq 0$, as well as their envelope, the singular solution $u_s \equiv 0$; cf. Example (E_{13}) of Chapter V. Since $f_{y'} = -2y'$ and $f_y \equiv 1$, for $f(t,y,y') = -y'^2 + y$, we see that the functions $\bar{u}_L$, $\bar{u}_R$ are unstable, while

$$f_{y'}[u_L(t)] = 2A^{1/2} - t \begin{cases} \geq 0 & \text{for } t \leq 2A^{1/2}, \\ < 0 & \text{for } t > 2A^{1/2} \end{cases},$$

and

$$f_{y'}[u_R(t)] = 1 - 2B^{1/2} - t \begin{cases} \leq 0 & \text{for } t \geq 1-2B^{1/2}, \\ > 0 & \text{for } t < 1-2B^{1/2}. \end{cases}$$

Consequently, u_L and u_R are strongly (weakly) stable if $A^{1/2} > 1/2$ $(A^{1/2} = 1/2)$ and $B^{1/2} > 1/2$ $(B^{1/2} = 1/2)$, respectively. If $0 < A^{1/2}, B^{1/2} < 1/2$ then these functions respectively lose stability at $t_L = 2A^{1/2}$ and $t_R = 1 - 2B^{1/2}$. Finally, $u_s \equiv 0$ is (I_0)-stable in $[0,1]$, since $f_y \equiv 1$.

Case 1. $A^{1/2}, B^{1/2} > 1/2$. Checking first for boundary layer behavior we see that since $f_{y'y'} < 0$, the inequalities required are $u_L(1) = 1/4(2A^{1/2} - 1)^2 \geq B$ and $u_R(0) = 1/4(2B^{1/2} - 1)^2 \geq A$. For these values Theorem 5.1 tells us that the problem has solutions such that

$$\lim_{\varepsilon \to 0^+} y(t,\varepsilon) = u_L(t) \quad \text{in } [0,1-\delta] \quad \text{if } 1/4(2A^{1/2} - 1)^2 \geq B$$

and

$$\lim_{\varepsilon\to 0^+} y(t,\varepsilon) = u_R(t) \quad \text{in} \quad [\delta,1] \quad \text{if} \quad 1/4(2B^{1/2} - 1)^2 \geq A. \tag{8.16}$$

Next, if $u_L(1) = 1/4(2A^{1/2} - 1)^2 < B$ and $u_R(0) = 1/4(2B^{1/2} - 1)^2 < A$, then it is easy to see that u_L intersects u_R angularly at $t_0 = A^{1/2} - B^{1/2} + 1/2$ in $(0,1)$. The existence of a solution satisfying

$$\lim_{\varepsilon\to 0^+} y(t,\varepsilon) = \begin{cases} u_L(t), & 0 \leq t \leq t_0, \\ u_R(t), & t_0 \leq t \leq 1, \end{cases} \tag{8.17}$$

follows from the theorem of Haber and Levinson.

<u>Case 2</u>. $A,B \leq 0$. For these values of A and B, there is no regular reduced solution which satisfies either of the boundary conditions, because $u = u'^2 \geq 0$. However, since $u_s \equiv 0$ is (I_0)-stable and $u_s(0) \geq A$, $u_s(1) \geq B$, we can apply Theorem 5.5 to conclude that the problem has a solution satisfying in $[0,1]$

$$A \exp[-t/\varepsilon^{1/2}] + B \exp[-(1-t)/\varepsilon^{1/2}] \leq y(t,\varepsilon) \leq 0.$$

<u>Case 3</u>. $B > 0$, $A \leq 0$. Here $u_R(t) = 1/4(t + 2B^{1/2} - 1)^2$ exists in $[0,1]$; however, there is no regular reduced solution satisfying the left-hand boundary condition. If $B^{1/2} > 1/2$ then u_R is strongly stable in $[0,1]$, and since $u_R(0) > A$ we obtain the limiting relation (8.16).

On the other hand, if $0 < B^{1/2} < 1/2$ then we know that u_R loses stability at $t_R = 1 - 2B^{1/2}$ in $(0,1)$, where it smoothly crosses the singular solution, that is, $u_R(t_R) = u_R'(t_R) = 0$. Since $A \leq 0$ Theorem 5.27 tells us that for such A and B the solution of the problem satisfies

$$\lim_{\varepsilon\to 0^+} y(t,\varepsilon) = \begin{cases} 0, & \delta \leq t \leq 1-2B^{1/2}, \\ u_R(t), & 1-2B^{1/2} \leq t \leq 1. \end{cases}$$

Finally, if $B^{1/2} = 1/2$ then $f_{y'}[u_R(t)] \leq 0$ for t in $[0,1]$; however, since $f_y \equiv 1$ Theorem 5.2 guarantees that the solution satisfies (8.16).

<u>Case 4</u>. $A > 0$, $B \leq 0$. This case is the reflection of Case 3, in that the statements made there apply with B, u_R and t replaced by A, u_L and $(1-t)$, respectively. We omit the details, except to note that for $0 < A^{1/2} < 1/2$, u_L intersects $u_s \equiv 0$ smoothly at the point $t_L = 2A^{1/2}$.

<u>Case 5</u>. $0 < A^{1/2}, B^{1/2} < 1/2$. For this last case, we note that u_L and u_R lose stability at $t_L = 2A^{1/2}$ and $t_R = 1-2B^{1/2}$, respectively, and

since $u_s(0) < A$ and $u_s(1) < B$, there can be no boundary layer behavior for these boundary values. We distinguish however two types of interior crossings:

(i) $A^{1/2} + B^{1/2} < 1/2$. In this case $0 < t_L < t_R < 1$, that is, $u_L(t_L) = u_L'(t_L) = 0$ and $u_R(t_R) = u_R'(t_R) = 0$, and so Theorem 5.25 implies that the solution satisfies

$$\lim_{\varepsilon\to 0^+} y(t,\varepsilon) = \begin{cases} u_L(t), & 0 \le t \le 2A^{1/2} \\ 0, & 2A^{1/2} \le t \le 1-2B^{1/2}, \\ u_R(t), & 1-2B^{1/2} \le t \le 1. \end{cases}$$

(ii) $A^{1/2} + B^{1/2} > 1/2$. Here $t_L > t_R$ and so u_L intersects u_R angularly at $t_0 = A^{1/2} - B^{1/2} + 1/2$ in $(0,1)$. The solution of the problem thus satisfies (8.17).

Finally, if $A^{1/2} + B^{1/2} = 1/2$ then $u_L \equiv u_R$, and the solution is uniformly close to this function in $[0,1]$ as $\varepsilon \to 0^+$.

Example 8.14. Consider next the problem

$$\varepsilon y'' = y'^2 - 2ty' + y, \quad -1 < t < 1,$$

$$y(-1,\varepsilon) = A, \quad y(1,\varepsilon) = B.$$

For the reduced equation $u = 2tu' - u'^2$ the parabola $u_p(t) = t^2$ is the p-discriminant locus and the t-axis $u_I \equiv 0$ is the inflection locus (cf. [48; Chapter 3], [38]). Clearly, u_I is a reduced solution, while u_p is not. Another solution which passes through the origin is the parabola $u_1(t) = 3/4t^2$, which is interesting in that $f_{y'}[u_1(t)] = t$ (for $f(t,y,y') = y'^2 - 2ty' + y$), and so u_1 is locally strongly stable in $[-1,1]$ (cf. Definition 5.4). The general parametric solution of the reduced equation is

$$t = 2/3p + cp^{-2},$$

$$u = 2tp - p^2 = 1/3p^2 + 2cp^{-1}, \quad p \ne 0, \ \sqrt[3]{c}.$$

Suppose first that $c > 0$.

Case (i). $-\infty < p < 0$. As $p \to -\infty$ we have $t \to -\infty$ and $u \to +\infty$; while as $p \to 0^-$ we have $t \to +\infty$ and $u \to -\infty$. These solutions are denoted by u_R.

Case (ii). $0 < p < \sqrt[3]{c}$. As $p \to 0^+$ we have $t \to +\infty$ and $u \to +\infty$; while as $p \to \sqrt[3]{c}^-$ we have $t \to t^*$ and $u \to t^{*2}$. These curves end on the p-discriminant locus and are denoted by u_R.

Case (iii). $\sqrt[3]{c} < p < \infty$. As $p \to \sqrt[3]{c}^{\,+}$ we have $t \to t^*$ and $u \to t^{*2}$; while as $p \to +\infty$ we have $t \to +\infty$ and $u \to +\infty$. These curves form cusps with the curves $\tilde{u}_R$ on u_p and lie between u_p and u_1. We denote them by $\hat{u}_R$. The curves in these last three cases are shown in Figure 8.4.

If $c < 0$ then the family of solution curves is obtained from the curves of Cases (i) - (iii) by reflection; we call the corresponding solutions u_L, $\tilde{u}_L$, $\hat{u}_L$. Their graphs are shown in Figure 8.5.

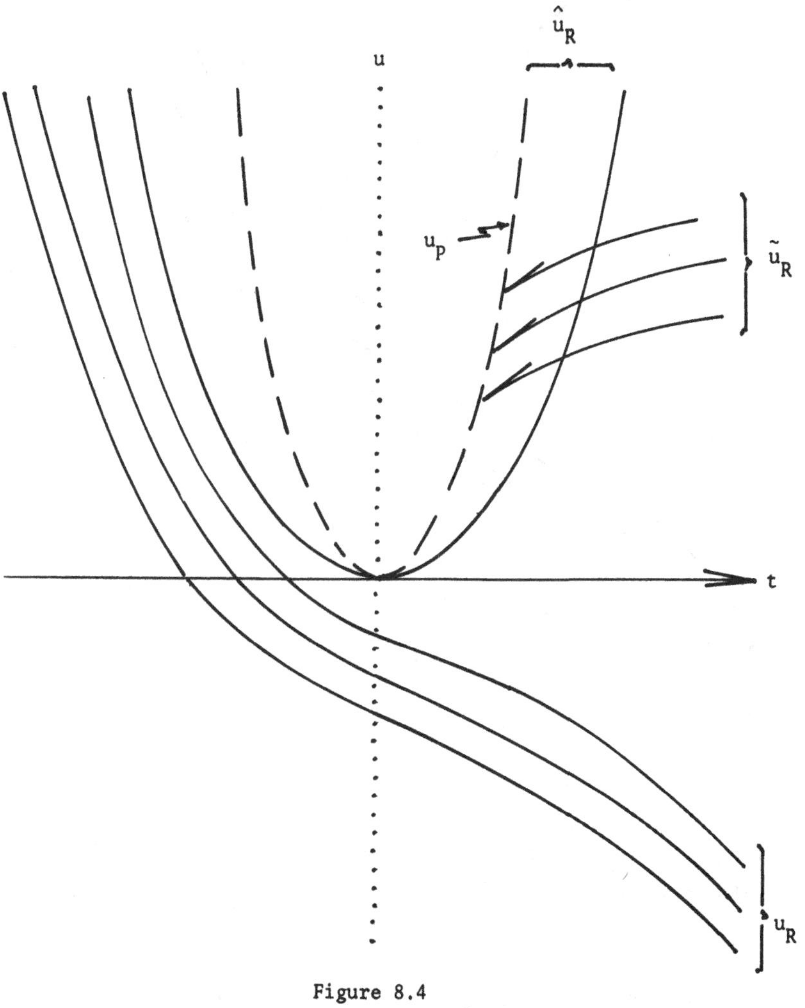

Figure 8.4

Solutions of $u = 2tu' - u'^2$ for $c > 0$.

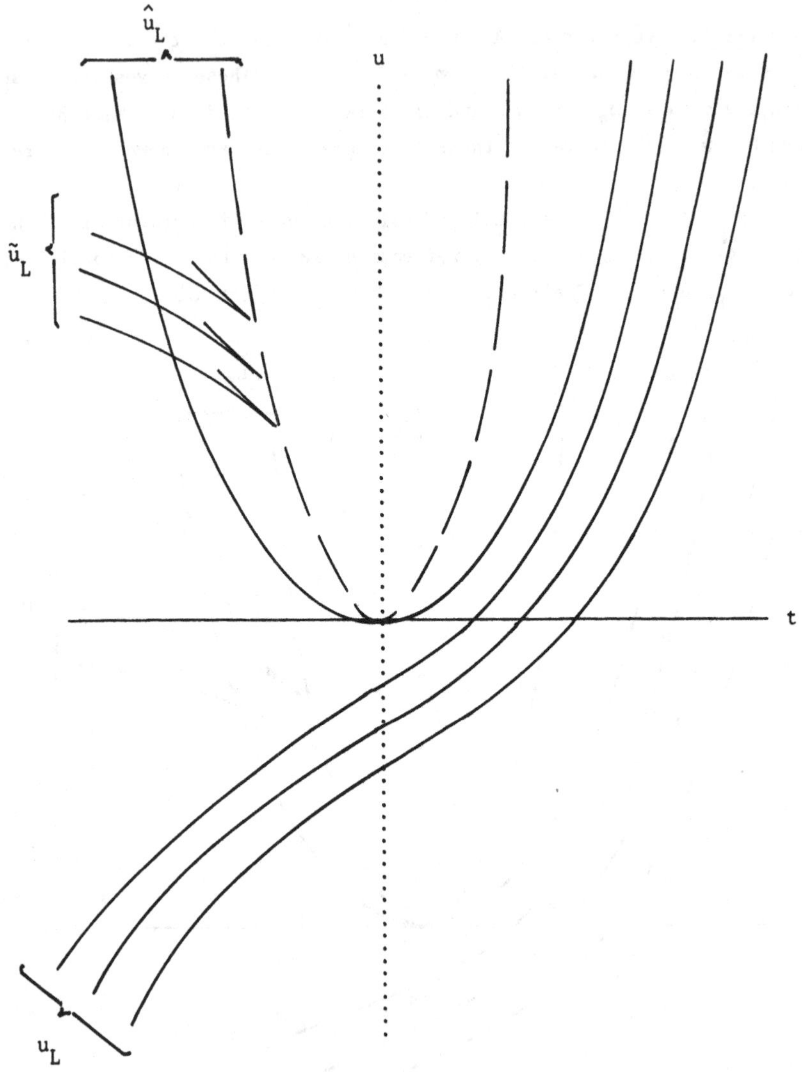

Figure 8.5
Solutions of $u = 2tu' - u'^2$ for $c < 0$.

Finally if $c = 0$ then we obtain the curve $u_1(t) = 3/4\ t^2$. The stability of these solutions is determined as follows:

$f_{y'}[u_R(t)] < 0$ for $-\infty < t < \infty$;

$f_{y'}[u_R(t)] \leq 0$, $f_{y'}[\hat{u}_R(t)] \geq 0$ for $t^* \leq t < \infty$;

$$f_{y'}[u_L(t)] > 0 \quad \text{for} \quad -\infty < t < \infty;$$

$$f_{y'}[\tilde{u}_L(t)] \geq 0, \quad f_{y'}[\hat{u}_L(t)] \leq 0, \quad \text{for} \quad -\infty < t \leq -t^*;$$

and finally

$$f_{y'}[u_1(t)] = t \quad \text{and} \quad f_{y'}[u_I] = -2t.$$

Thus the solutions u_L and u_R are strongly stable, $\tilde{u}_L$ and $\tilde{u}_R$ are weakly stable and u_1 is locally strongly stable, while $\hat{u}_L$ and $\hat{u}_R$ are unstable. We now investigate the behavior of the solution $y = y(t,\varepsilon)$ of the problem as $\varepsilon \to 0^+$ by fixing A and varying B.

<u>Case 1</u>. $A > 3/4$. If $B \geq 3/4$ then since u_1 is locally strongly stable and (I_0)-stable, Theorem 5.5 tells us that there are boundary layers at both endpoints, that is,

$$\lim_{\varepsilon\to 0^+} y(t,\varepsilon) = 3/4\, t^2 \quad \text{in} \quad [-1+\delta, 1-\delta].$$

Next, for $0 < B < 3/4$ there is a reduced solution $\tilde{u}_R$ such that $\tilde{u}_R(1) = B$ and $f_{y'}[\tilde{u}_R(t)] < 0$ in $(t^*,1]$. For such A, B there is an angular crossing between u_1 and $\tilde{u}_R$ at a point t_2 in $(t^*,1)$. Since $f_{y'}[u_1(t)] > 0$ in $(0,1]$ we conclude from Theorem 5.27 that there is an angular layer at t_2 and a boundary layer at $t = -1$, that is,

$$\lim_{\varepsilon\to 0^+} y(t,\varepsilon) = \begin{cases} 3/4\, t^2, & -1+\delta \leq t \leq t_2, \\ u_R(t), & t_2 \leq t \leq 1. \end{cases}$$

If now $B = 0$ then the reduced path

$$u_0(t) = \begin{cases} u_1(t), & -1 \leq t \leq 0, \\ 0, & 0 \leq t \leq 1, \end{cases}$$

is continuously differentiable, locally weakly stable $(f_{y'}[u_0(t)] \leq 0)$, and (I_0)-stable. Theorem 5.5 implies that

$$\lim_{\varepsilon\to 0^+} y(t,\varepsilon) = u_0(t) \quad \text{in} \quad [-1+\delta, 1].$$

Finally, if $B < 0$ then there is a strongly stable reduced solution u_R such that $u_R(1) = B$ and $u_R(-1) < A$. Theorem 5.1 implies that

$$\lim_{\varepsilon\to 0^+} y(t,\varepsilon) = u_R(t) \quad \text{in} \quad [-1+\delta, 1]. \tag{8.18}$$

Case 2. $A = 3/4$. The same results as in Case 1 are valid with the exception that there is no boundary layer at $t = -1$ if $B \geq 0$.

Case 3. $0 < A < 3/4$. For this range of A there is a reduced solution $\tilde{u}_L$ such that $\tilde{u}_L(-1) = A$, with $f_{y'}[u_L(t)] \geq 0$ for $-1 \leq t \leq -t^*$. If $B \geq 3/4$ then $\tilde{u}_L$ intersects u_1 at a point t_1 in $(-1,-t^*)$, and there is an angular layer there. Since $u_1(1) \leq B$ there is also a boundary layer at $t = 1$, that is,

$$\lim_{\varepsilon\to 0^+} y(t,\varepsilon) = \begin{cases} u_L(t), & -1 \leq t \leq t_1, \\ u_1(t), & t_1 \leq t \leq 1-\delta. \end{cases}$$

If $0 < B < 3/4$ then there is an angular crossing between $\tilde{u}_L$ and u_1 at a point t_1 in $(-1,-t^*)$ and an angular crossing between u_1 and $\tilde{u}_R$ at a point t_2 in $(t^*,1)$. We conclude from Theorem 5.25 that

$$\lim_{\varepsilon\to 0^+} y(t,\varepsilon) = \begin{cases} \tilde{u}_L(t), & -1 \leq t \leq t_1, \\ u_1(t), & t_1 \leq t \leq t_2, \\ \tilde{u}_R(t), & t_2 \leq t \leq 1. \end{cases}$$

If $B = 0$, then $\tilde{u}_L$ intersects u_0 at a point t_1 in $(-1,-t^*)$, and so

$$\lim_{\varepsilon\to 0^+} y(t,\varepsilon) = \begin{cases} \tilde{u}_L(t), & -1 \leq t \leq t_1, \\ u_0(t), & t_1 \leq t \leq 1. \end{cases}$$

Finally, for $B < 0$ the solution is described by the limiting relation (8.18), if $u_R(-1) \leq A$. Otherwise, if $u_R(-1) > A$ then $\tilde{u}_L$ intersects u_R at a point t_1 in $(-1,-t^*)$, and so

$$\lim_{\varepsilon\to 0^+} y(t,\varepsilon) = \begin{cases} \tilde{u}_L(t), & -1 \leq t \leq t_1, \\ u_R(t), & t_1 \leq t \leq 1. \end{cases}$$

Case 4. $A = 0$. The same results as in Case 3 obtain, with $\tilde{u}_L$ replaced by $u_I \equiv 0$ and with the important exception that u_I intersects u_1 smoothly at $t = 0$, that is, $u_I(0) = u_1(0) = u_I'(0) = u_1'(0) = 0$.

Case 5. $A < 0$. This is the final case. For such A there is a strongly stable reduced solution u_L such that $u_L(-1) = A$. If $B \geq 3/4$ then

$$\lim_{\varepsilon\to 0^+} y(t,\varepsilon) = u_L(t) \quad \text{in} \quad [-1,1-\delta], \tag{8.19}$$

since $u_L(1) < B$, by virtue of Theorem 5.1. Again, by the same theorem if $0 < B < 3/4$ and $u_L(1) \leq B$, then (8.19) is valid; whereas, if $u_L(1) > B$, then u_L intersects $\tilde{u}_R$ at a point t_2 in $(t^*,1)$, and so

$$\lim_{\varepsilon \to 0^+} y(t,\varepsilon) = \begin{cases} u_L(t), & -1 \leq t \leq t_2, \\ \tilde{u}_R(t), & t_2 \leq t \leq 1. \end{cases} \tag{8.20}$$

If $B = 0$, then again (8.19) is valid, if $u_L(1) \leq 0$; whereas, for $u_L(1) > 0$, u_L intersects $u_I \equiv 0$ at a point t_2 in $(t^*,1)$, and so (8.20) is valid with $\tilde{u}_R$ replaced by 0. Finally, for $B < 0$ the relation (8.19) holds if $u_L(1) \leq B$, while (8.18) holds if $u_R(-1) \leq A$. Otherwise, if $u_L(1) > B$ and $u_R(-1) > A$, then u_L intersects u_R at a point t_0 in $(-1,1)$, and so by Haber and Levinson's theorem

$$\lim_{\varepsilon \to 0^+} y(t,\varepsilon) = \begin{cases} u_L(t), & -1 \leq t \leq t_0, \\ u_R(t), & t_0 \leq t \leq 1. \end{cases}$$

Example 8.15. Consider now the Robin problem

$$\varepsilon y'' = -y'^2 + y, \quad 0 < t < 1,$$

$$y(0,\varepsilon) = y'(0,\varepsilon) = A, \quad y(1,\varepsilon) = B,$$

for various values of A and B. It follows directly from the "Robin" version of Lemma 2.1 that this problem has for each A, B a unique solution $y = y(t,\varepsilon)$ for all $\varepsilon > 0$ which satisfies in $[0,1]$

$$\min\{A,B,0\} \leq y(t,\varepsilon) \leq \max\{A,B,0\}.$$

The differential equation is the one discussed in Example 8.13, and so we know that the reduced equation has the solutions $u(t) = 1/4(t+c)^2$, their envelope $u_s \equiv 0$, and paths formed from finite combinations of these curves, for example,

$$u(t) = \begin{cases} 0, & 0 \leq t \leq c < 1, \\ 1/4(t-c)^2, & c \leq t \leq 1. \end{cases}$$

Clearly the reduced problems

$$u = u'^2, \quad u(0) - u'(0) = A, \tag{R_L}$$

and

$$u = u'^2, \quad u(1) = B, \tag{R_R}$$

each have two solutions: $u_L(t) = 1/4(t + 1 - (1+4A)^{1/2})^2$, $\tilde{u}_L(t) = 1/4(t + 1 + (1+4A)^{1/2})^2$ (if $A \geq -1/4$) and $u_R(t) = 1/4(t + 2B^{1/2} - 1)^2$,

$\tilde{u}_R(t) = 1/4(t - 2B^{1/2} - 1)^2$ (if $B \geq 0$). Of these, $\tilde{u}_L$ and $\tilde{u}_R$ are unstable, while u_L is strongly stable in $[0,(1+4A)^{1/2} - 1)$ if $A > 0$ and u_R is strongly stable in $(1-2B^{1/2},1]$ if $B > 0$. Note that u_L and u_R are also (I_0)-stable.

Let us first assume that $A > 3/4$. Then u_L is strongly stable in $[0,1]$, and so if $u_L(1) \geq B$, that is, if $B \leq 0$ or $B^{1/2} \leq 1/2(1+4A)^{1/2} - 1$, then the solution satisfies

$$\lim_{\varepsilon\to 0^+} y(t,\varepsilon) = u_L(t) \quad \text{in} \quad [0,1-\delta],$$

by virtue of Theorem 5.7. Next, if $B^{1/2} > 1/2$ then u_R is also strongly stable in $[0,1]$. We consider for what values of A the solution satisfies

$$\lim_{\varepsilon\to 0^+} y(t,\varepsilon) = u_R(t) \quad \text{in} \quad [0,1]. \tag{8.21}$$

The relation (†) of Theorem 5.9 clearly holds if $u_R(0) - u_R'(0) \geq A$, that is, if

$$B^{1/2} \leq 1 - 1/2(1+4A)^{1/2} \quad \text{or if} \quad B^{1/2} \geq 1 + 1/2(1+4A)^{1/2},$$

and so for such values of A, (8.21) follows. Suppose, however, that $u_R(0) - u_R'(0) < A$, that is,

$$1 - 1/2(1+4A)^{1/2} < B^{1/2} < 1 + 1/2(1+4A)^{1/2}. \tag{8.22}$$

Then it is still possible to satisfy (†) if, in addition, A is such that

$$-(u_R(0) - A)^2 + u_R(0) > 0,$$

that is, if

$$B^{1/2} > 1/2(1+4A)^{1/2} \quad \text{or if} \quad B^{1/2} < -1/2(1+4A)^{1/2}.$$

Clearly the latter inequality is incompatible with (8.22), and so we see that (†) is satisfied if

$$1/2(1+4A)^{1/2} < B^{1/2} < 1 + 1/2(1+4A)^{1/2}.$$

Thus, for this range of A, the relation (8.21) holds.

We consider next an application of the "(RP_5)" version of Theorem 5.30. Clearly u_L intersects u_R at the point $t_0 = 1/2((1+4A)^{1/2} - 2B^{1/2})$ in $(0,1)$ if and only if

$$u_L(1) < B, \quad u_R(0) - u_R'(0) < A \quad \text{and} \quad -(u_R(0)-A)^2 + u_R(0) < 0,$$

that is,

$$1/2(1+4A)^{1/2} - 1 < B < 1/2(1+4A)^{1/2}.$$

For such values the solution satisfies

$$\lim_{\varepsilon\to 0^+} y(t,\varepsilon) = \begin{cases} u_L(t), & 0 \le t \le t_0, \\ u_R(t), & t_0 \le t \le 1. \end{cases}$$

Our discussion has yet to involve the singular solution $u_s \equiv 0$. Since u_s is weakly stable and (I_0)-stable, we apply Theorem 5.5 if $A,B \le 0$ to see that

$$\lim_{\varepsilon\to 0^+} y(t,\varepsilon) = 0 \quad \text{in} \quad [0,1-\delta].$$

We also illustrate a smooth crossing. If $0 < A < 3/4$ then $1 < (1+4A)^{1/2} < 2$, and so u_L intersects u_s smoothly at $t_L = (1+4A)^{1/2} - 1$, that is, $u_L(t_L) = u_L'(t_L) = 0$. Thus for $B = 0$ the solution satisfies

$$\lim_{\varepsilon\to 0^+} y(t,\varepsilon) = \begin{cases} u_L(t), & 0 \le t \le t_L, \\ 0 \quad , & t_L \le t \le 1, \end{cases}$$

by the analog of Theorem 5.31. Similarly, if $0 < B < 1/4$, then u_R intersects u_s smoothly at $t_R = 1-2B^{1/2}$, and so for $A \le 0$

$$\lim_{\varepsilon\to 0^+} y(t,\varepsilon) = \begin{cases} 0, & 0 \le t \le t_R, \\ u_R(t), & t_R \le t \le 1. \end{cases}$$

Next, if $0 < A < 3/4$ and $0 < B < 1/4$, then there are three possibilities:

(i) u_L intersects u_s, and then u_s intersects u_R;

(ii) u_L and u_R intersect u_s at the same point, that is, $u_L \equiv u_R$;

(iii) u_L intersects u_R.

Clearly (i), (ii) and (iii) obtain if and only if $t_L < t_R$, $t_L = t_R$ and $t_L > t_R$, respectively.

Finally, for $0 < A < 3/4$ and $B < 0$ we know that u_L intersects u_s smoothly at t_L and $u_s > B$. Consequently the analog of Theorem 5.31 tells us that

$$\lim_{\varepsilon\to 0^+} y(t,\varepsilon) = \begin{cases} u_L(t), & 0 \le t \le t_L, \\ 0, & t_L \le t \le 1-\delta. \end{cases}$$

Application 8.5. We conclude this section with an application of the theory to the model problem (cf. [95] and Application 8.2)

$$\varepsilon y'' = \varepsilon\theta(1+\theta y)^{-1}y'^2 + y^r(1+\theta y), \quad 0 < t < 1,$$

$$-y'(0,\varepsilon) = 0, \quad y(1,\varepsilon) + \Sigma y'(1,\varepsilon) = 1,$$

which describes a chemical reaction accompanied by a change in volume. More precisely, y represents the dimensionless concentration of a gas A undergoing an isothermal reaction on a flat plate catalytic surface, namely $A \to Bn$. The variable t is the normalized distance from a plane of symmetry to the edge of the plate ($t = 1$), and the positive constant n is the stoichiometric factor of the reaction. As before, the parameter ε is the reciprocal of the square of the Thiele modulus, and the parameter $\theta \approx n-1$ is the volume change modulus. (Note that $\theta = 0$ or $n = 1$ in the absence of a volume change; cf. Application 8.2.) For simplicity, the reaction is assumed to be of integral order $r \geq 1$ and Σ is again the reciprocal of the Sherwood number.

If $n > 1$, then $\theta > 0$, and there is an increase in the volume due to the reaction. The coefficient of y'^2 in the differential equation is positive, and we can apply the quadratic theory if ε is sufficiently small. Since y is a concentration we are only interested in positive solutions, and we see immediately that $u_s \equiv 0$ is the only nonnegative solution of the reduced equation. Since u_s is stable in the sense of Definition 5.3 it follows from Theorem 5.19 that for positive values of Σ the problem has a solution $y = y(t,\varepsilon)$ satisfying in $[0,1]$

$$0 \leq y(t,\varepsilon) \leq \Sigma^{-1}\varepsilon^{1/2}\exp[-(1-t)/\varepsilon^{1/2}], \text{ if } r = 1,$$

and

$$0 \leq y(t,\varepsilon) \leq \Sigma^{-1}r\rho^{-1}\varepsilon^{1/(2r+2)}(1+\rho(1-t)/\varepsilon^{1/(2r+2)})^{-1/r}, \text{ if } r \geq 2,$$

where $\rho = \rho(r)$ is a known positive constant.

In conclusion, we note that if $\Sigma = 0$ (that is, if there is no resistance to the transfer of reactant from the bulk flow to the surface at $t = 1$), then Theorem 5.13 provides us with the estimates

$$0 \leq y(t,\varepsilon) \leq \exp[-(1-t)/\varepsilon^{1/2}], \text{ if } r = 1,$$

and

$$0 \leq y(t,\varepsilon) \leq (1+\rho_1(1-t)/\varepsilon^{1/2})^{-1/r}, \text{ if } r \geq 2,$$

where $\rho_1 = \rho_1(r)$ is again a known positive constant.

8.4. Examples of Superquadratic Problems and An Application

Example 8.16. Consider the differential equation

$$\varepsilon y'' = (y-t^2)^{2q+1}(1+y'^2)^{3/2} \equiv F(t,y,y'),$$

for $0 < t < 1$ and integral values of $q \geq 0$. The function $u(t) = t^2$ is the only real zero of F, and u is clearly (I_q)-stable in the sense of Definition 6.1. Since $u(0) = 0$, $u(1) = 1$ and $u'' \equiv 2$, the Dirichlet problem (cf. Example 8.1).

$$\varepsilon y'' = F(t,y,y'), \quad 0 < t < 1,$$

$$y(0,\varepsilon) = 0, \quad y(1,\varepsilon) = 1,$$

has by virtue of Theorem 6.1 a solution $y = y(t,\varepsilon)$ as $\varepsilon \to 0^+$ satisfying in $[0,1]$

$$t^2 \leq y(t,\varepsilon) \leq t^2 + (2\varepsilon)^{1/(2q+1)}.$$

On the other hand, since $u'(0) = 0$ Theorem 6.3 tells us that the Robin problem (cf. Example 8.5)

$$\varepsilon y'' = F(t,y,y'), \quad 0 < t < 1,$$

$$-y'(0,\varepsilon) = A, \quad y(1,\varepsilon) = 1,$$

has a solution as $\varepsilon \to 0^+$ satisfying in $[0,1]$

$$t^2 \leq y(t,\varepsilon) \leq t^2 + v(t,\varepsilon) + (2\varepsilon)^{1/(2q+1)},$$

for any value of $A \geq 0$. Here $v(t,\varepsilon) \equiv 0$ if $A = 0$, and if $A > 0$, then

$$v(t,\varepsilon) = A\varepsilon^{1/2}\exp[-t/\varepsilon^{1/2}] \quad (q = 0)$$

and

$$v(t,\varepsilon) = A\tau^{-1}q\varepsilon^{1/(2q+2)}(1 + \tau t/\varepsilon^{1/(2q+2)})^{-1/q} \quad (q \geq 1),$$

for

$$\tau = \tau(q) = [A^{2q}q^{2q+2}/(q+1)]^{1/(2q+2)}.$$

Application 8.6. The superquadratic theory is often useful in solving problems involving the curvature of surfaces, since nonlinearities of the form $(1 + y'^2)^{3/2} \sim |y'|^3$, as $|y'| \to \infty$, arise naturally. As an illustration, we note (cf. [3; Chapter 1] or [78]) that the elevation y of the free surface of a liquid meeting a plane, vertical rigid wall (at $t = 0$) is described by the problem

$$\varepsilon y'' = y(1 + y'^2)^{3/2}, \quad 0 < t < L,$$

$$y'(0,\varepsilon) = \tan\theta, \quad |\theta| < \pi/2, \; y(L,\varepsilon) = O(\varepsilon),$$

for $\varepsilon^2 = T/(\rho g)$ and L an arbitrarily large positive constant. Here ρ is the density of the liquid, g is the gravitational constant, T is the coefficient of surface tension, and θ is the contact angle, the angle the surface makes with the wall, measured from the horizontal axis, $y = 0$. If T is small then ε is small, and the problem for the elevation is singularly perturbed.

Suppose first that the liquid is water, and therefore, that $0 \leq \theta < \pi/2$. Since $u \equiv 0$ is an (I_0)-stable reduced solution and $-u'(0) = 0 \leq \tan\theta$, Theorem 6.3 tells us that the elevation y satisfies in $[0,L]$

$$0 \leq y(t,\varepsilon) \leq (\tan\theta)\varepsilon^{1/2}\exp[-t/\varepsilon^{1/2}] + O(\varepsilon).$$

However, if the liquid is mercury, then the contact angle is negative $(-\pi/2 < \theta < 0)$, Since $-u'(0) = 0 > \tan\theta$ we now conclude that the "depression" y satisfies in $[0,L]$

$$O(\varepsilon) + (\tan\theta)\varepsilon^{1/2}\exp[-t/\varepsilon^{1/2}] \leq y(t,\varepsilon) \leq 0.$$

Example 8.17. Consider now the problem

$$\varepsilon y'' = -y' - y'^3, \quad 0 < t < 1,$$

$$py(0,\varepsilon) - y'(0,\varepsilon) = A, \quad y(1,\varepsilon) = B, \quad p \geq 0.$$

(We saw earlier that the Dirichlet problem $\varepsilon y'' = -y' - y'^3$, $y(0,\varepsilon) = A$, $y(1,\varepsilon) = B$, has no solution as $\varepsilon \to 0^+$ if $A \neq B$.) The reduced equation $f(u') = -u' - u'^3$ has $u' = 0$ as its only real solution, and since $f_{u'}(0) = -1$ we make the corresponding reduced solution satisfy $u(1) = B$, that is, $u = u_R(t) \equiv B$. Suppose first that $p = 0$. If $A = 0$ then $y(t,\varepsilon) \equiv B$ is the solution. However, if $A \neq 0$ then $Af(\lambda) = -A\lambda(1+\lambda^2) > 0$ for all values of λ between 0 and $-A$, $\lambda \neq 0$. Consequently Theorem 6.11 tells us that for all values of A the solution satisfies in $[0,1]$

$$y(t,\varepsilon) = B + O(\varepsilon|A|\exp[-t/\varepsilon]).$$

Suppose finally that $p > 0$. If $A = pB$ then $y(t,\varepsilon) \equiv B$ is the solution, while if $A \neq pB$ then

$$(pB - A)f(\lambda) = -\lambda(pB - A)(1 + \lambda^2) < 0,$$

for all values of λ between 0 and $pB - A$, $\lambda \neq 0$. Again from Theorem 6.11 we see that for all values of A and B

$$y(t,\varepsilon) = B + O(\varepsilon|pB-A|\exp[-t/\varepsilon]) \quad \text{in} \quad [0,1].$$

Example 8.18. Consider next the related problem

$$\varepsilon y'' = y' - y'^3, \quad 0 < t < 1,$$

$$y(0,\varepsilon) - y'(0,\varepsilon) = A, \quad y(1,\varepsilon) + y'(1,\varepsilon) = B.$$

The reduced equation $f(u') = u' - u'^3 = 0$ now has three solutions $u_1' = 1$, $u_2' = -1$ and $u_3' = 0$, which satisfy $f_{u'}(\pm 1) = -2$ and $f_{u'}(0) = 1$. Thus we make u_1 and u_2 satisfy $u_j(1) + u_j'(1) = B$ for $j = 1,2$, that is, $u_1(t) = t + B - 2$ and $u_2(t) = -t + B + 2$, and we make u_3 satisfy $u_3(0) - u_3'(0) = A$, that is, $u_3 \equiv A$. Consider u_1 first. If $A = B - 3$ then $u_1(0) - u_1'(0) = A$, and so $y(t,\varepsilon) = t + B - 2$ is the solution. However, if $A < B - 3$ then $(u_1(0) - 1 - A)f(\lambda) = (B - 3 - A)\lambda(1-\lambda^2) < 0$ for all values of λ between 1 and $B - 2 - A$, $\lambda \neq 1$. Theorem 6.11 implies that the solution satisfies in $[0,1]$

$$y(t,\varepsilon) = u_1(t) + O(\tfrac{1}{2}\,\varepsilon(B-3-A)\exp[-2t/\varepsilon]). \tag{8.23}$$

Finally, if $A > B - 3$ then $(B - 3 - A)\lambda(1 - \lambda^2) > 0$ for λ between 1 and $B - 2 - A$, $\lambda \neq 1$, *provided* that $B - 2 - A > 0$. Again from Theorem 6.11 it follows that for $B - 3 < A < B - 2$ the solution satisfies in $[0,1]$

$$y(t,\varepsilon) = u_1(t) + O((\varepsilon/k)|B-3-A|\exp[-kt/\varepsilon]), \tag{8.24}$$

for a positive constant $k < 2$.

The asymptotic behavior described by u_2 is clearly a reflection of that described by u_1. In particular, if $B + 3 \leq A$ $[B + 2 < A < B + 3]$ then the relation (8.23) [(8.24)] obtains with u_1 replaced by u_2 and the term $|B - 3 - A|$ replaced by $|B + 3 - A|$ inside the Landau symbols.

Consider next the reduced solution $u_3 \equiv A$. If $A = B$ then $y(t,\varepsilon) \equiv A$ is the solution, while if $A < B$ then $(A-B)f(\lambda) = (A-B)\lambda(1 - \lambda^2) < 0$ for λ between 0 and $B - A$, $\lambda \neq 0$, provided $B - A < 1$. Similarly, if $A > B$ then this inequality obtains provided $B - A > -1$. Consequently for $B - 1 < A < B + 1$ Theorem 6.11 tells us that the solution satisfies in $[0,1]$

$$y(t,\varepsilon) = A + O((\varepsilon/k)|B - A|\exp[-kt/\varepsilon]),$$

for a positive constant $k < 1$.

Summarizing to this point, we have estimates of the solution for all values of A and B, except those satisfying $B - 2 \le A \le B - 1$ and $B + 1 \le A \le B + 2$, for which Theorem 6.11 is inapplicable. Thus we are led to consider the angular paths

$$u_4(t) = \begin{cases} A, & 0 \le t \le t_0, \\ t+B-2, & t_0 \le t \le 1, \end{cases} \qquad u_5(t) = \begin{cases} A, & 0 \le t \le \tilde{t}_0, \\ -t+B+2, & \tilde{t}_0 \le t \le 1. \end{cases}$$

It follows directly that $t_0 = A - B + 2$ is in $(0,1)$ if and only if $B - 2 < A < B - 1$, while $\tilde{t}_0 = B - A + 2$ is in $(0,1)$ if and only if $B + 1 < A < B + 2$. Consider just u_4. For $\sigma_L = 0$ and $\sigma_R = 1$ we see that $(\sigma_R - \sigma_L)f(\lambda) = \lambda(1-\lambda^2) > 0$ for $0 < \lambda < 1$, and so Theorem 6.12 implies that for $B - 2 < A < B - 1$

$$y(t,\varepsilon) = u_4(t) + O(\tfrac{1}{2}(\varepsilon/k)\exp[-k|t - t_0|/\varepsilon])$$

in $[0,1]$, for a positive constant $k < 1$. A similar result holds for $B + 1 < A < B + 2$ with u_4, t_0 replaced by u_5, $\tilde{t}_0$.

Finally, if $A = B - 2$ $[A = B + 2]$ then $\lim_{\varepsilon\to0^+} y(t,\varepsilon) = t + B - 2$ $[-t + B + 2]$, while if $A = B - 1$ or $A = B + 1$, then $\lim_{\varepsilon\to0^+} y(t,\varepsilon) = 0$. None of these limits is a surprise; the convergence is, of course, uniform in $[0,1]$.

Example 8.19. The solutions of the nonautonomous problem

$$\varepsilon y'' = y - ty' - y'^3 \equiv f(t,y,y'), \quad -1 < t < 1,$$

$$-y'(-1,\varepsilon) = A, \quad y(1,\varepsilon) = B,$$

exhibit the types of boundary and interior layer behavior encountered in the two previous examples and more, for various choices of A and B. Here the reduced equation $u = tu' + u'^3$ is a Clairaut equation whose solutions are the one-parameter family of straight lines $u(t) = ct + c^3$ and their envelope, the singular solution $u_s(t) = \pm 2(-t)^{3/2}/(3\sqrt{3})$, defined for $t \le 0$.

Suppose first that $B = 2$. Then $u_R(t) = t + 1$ is a solution of the reduced problem (R_R) which is stable since $f_{y'}[u_R(t)] = -t - 3 \le -2$ in $[-1,1]$. In order to apply Theorem 6.11 we must determine for what values of A

$$(-u_R'(-1) - A)f(-1,u_R(-1),\lambda) = (-1-A)\lambda(-1-\lambda^2) < 0 \tag{8.25}$$

for λ between $u_R'(-1) = 1$ and $-A$, $\lambda \neq 1$. If $A > -1$ then (8.25) obtains provided $A < 0$, while it clearly obtains for all $A < -1$. Finally, if $A = -1$ then $y(t,\varepsilon) = u_R(t)$ is the solution. We conclude that for all $A < 0$ the solution satisfies in $[-1,1]$

$$\lim_{\varepsilon\to 0^+} y(t,\varepsilon) = t + 1.$$

Suppose next that $A = 0$ and $B = 5/8$. Then $u_L \equiv 0$ is a solution of the reduced problem (R_L) and $u_R(t) = \frac{1}{2}t + \frac{1}{8}$ is a solution of (R_R) which intersect at $t_0 = -1/4$. The corresponding angular path $u_1(t) = u_L(t)$ in $[-1,-1/4]$, $u_1(t) = u_R(t)$ in $[-1/4,1]$ is globally stable since $f_{y'}[u_L(t)] = -t \geq \frac{1}{4}$ in $[-1,-\frac{1}{4}]$ and $f_{y'}[u_R(t)] = -t - \frac{3}{4} \leq -\frac{1}{2}$ in $[-\frac{1}{4}, 1]$. To apply Theorem 6.12 we note that

$$(\sigma_R - \sigma_L)f(-\tfrac{1}{4}, 0,\lambda) > 0$$

for λ between 0 and $\frac{1}{2}$. Consequently the solution satisfies in $[-1,1]$

$$\lim_{\varepsilon\to 0^+} y(t,\varepsilon) = u_1(t).$$

We consider finally two applications of the "singular" analogs of Theorems 6.11 and 6.12. Set $B = 10/27$. The reduced problem (R_R) has the solution $u_R(t) = t/3 + 1/27$, and it intersects the lower branch u_s $(= -2(-t)^{3/2}/(3\sqrt{3}))$ of the singular solution at the point $t_2 = -1/3$. Since u_s is singular $u_s'(t_2) = u_R'(t_2)$, and therefore the reduced path $u_3(t) = u_s(t)$ in $[-1,-1/3]$, $u_3(t) = u_R(t)$ in $[-1/3,1]$ is continuously differentiable. It remains for us to determine the values of A for which

$$(-u_s'(-1) - A)f(-1,u_s(-1),\lambda) < 0$$

for λ between $1/\sqrt{3}$ and $-A$, $\lambda \neq 1/\sqrt{3}$. A short calculation shows that this inequality obtains for all $A < -1/\sqrt{3}$. We conclude that the solution satisfies

$$\lim_{\varepsilon\to 0^+} y(t,\varepsilon) = u_3(t) = \quad [-1,1].$$

(If $A = -1/\sqrt{3}$, then $y(t,\varepsilon) = u_3(t)$.) Finally, let $A = 1/3$ and $B = 26/27$. Then $u_R(t) = 2t/3 + 8/27$ intersects the upper branch u_s $(= 2(-t)^{3/2}/(3\sqrt{3}))$ of the singular solution at $t_2 = -1/3$. However, for this choice of A and B, $\sigma_L = u_s'(-1/3) = -1/3 < 2/3 = u_R'(-1/3) = \sigma_R$, in contrast to the previous case. A short calculation shows that

$$(\sigma_R - \sigma_L)f(-1/3,2/27,\lambda) > 0$$

for $-1/3 < \lambda < 2/3$, and so we conclude that

$$\lim_{\varepsilon\to 0^+} y(t,\varepsilon) = \begin{cases} u_s(t), & -1 \le t \le -1/3, \\ u_R(t), & -1/3 \le t \le 1. \end{cases}$$

Example 8.20. We close this section with the problem

$$\varepsilon y'' = y + ty' + y'^n \equiv f(t,y,y'), \quad -1 < t < 1,$$

$$y(-1,\varepsilon) - y'(-1,\varepsilon) = A, \quad y(1,\varepsilon) + y'(1,\varepsilon) = B,$$

for $n \ge 3$ an integer. The function $u \equiv 0$ is clearly (I_0)-stable in the sense of Definition 6.1, and it is also locally strongly y'-stable since $f_{y'}[0] = t$. Suppose first that n is odd. In order to proceed we consider the two inequalities

$$(u(-1) - u'(-1) - A)f(-1, u(-1),\lambda) < 0,$$

for λ between 0 and $-A$, $\lambda \neq 0$, and

$$(u(1) + u'(1) - B)f(1,u(1),\lambda) < 0,$$

for λ between 0 and B, $\lambda \neq 0$. The first inequality clearly obtains for $|A| < 1$, $A \neq 0$, since n is odd, while the second obtains for all values of $B \neq 0$. If $A = B = 0$ then $y(t,\varepsilon) \equiv 0$, and so we deduce by arguing as in the proof of Theorem 6.11 that if n is odd and $|A| < 1$, then for all values of B the solution satisfies

$$\lim_{\varepsilon\to 0^+} y(t,\varepsilon) = 0 \text{ in } [-1,1]. \tag{8.26}$$

On the other hand, if n is even then these inequalities obtain for all values of $A > -1$, $A \neq 0$, and $B > -1$, $B \neq 0$, respectively. We again have the limiting relation (8.26).

Part II - VECTOR PROBLEMS

§8.5. Examples of Semilinear Systems and An Application

Example 8.21. Let us illustrate the norm-bound theory of Chapter VII by first considering the two-dimensional system in $(0,1)$

$$\varepsilon y_1'' = y_1-y_2-y_1^3 \equiv h_1(y_1,y_2), y_1(0,\varepsilon) = A_1, y_1(1,\varepsilon) = B_1,$$

$$\varepsilon y_2'' = y_2+y_1-y_2^3 \equiv h_2(y_1,y_2), y_2(0,\varepsilon) = A_2, y_2(1,\varepsilon) = B_2.$$

The corresponding reduced system $\underset{\sim}{h} = (h_1\ h_2)^T = (0\ 0)^T$ has the solution $\underset{\sim}{u} = \underset{\sim}{0}$, and it is clearly stable, in the sense of Definition 7.1, since the matrix $J(0,0)$ is positive definite, for $J(\underset{\sim}{y}) = \begin{pmatrix} 1-3y_1^2 & -1 \\ 1 & 1-3y_2^2 \end{pmatrix}$ the Jacobian matrix. Finally, the quadratic form $\underset{\sim}{y}^T J(\underset{\sim}{y})\underset{\sim}{y} = y_1^2(1-3y_1^2) + y_2^2(1-3y_2^2)$ is positive definite, only for vectors $(y_1\ y_2)^T$ satisfying $|y_1| < 1/\sqrt{3}$ and $|y_2| < 1/\sqrt{3}$. Consequently, for boundary values A_1, A_2 and B_1, B_2 satisfying $|A_i| < 1/\sqrt{3}$, $|B_i| < 1/\sqrt{3}$ $(i = 1,2)$ Theorem 7.1 tells us that the problem has a solution $\underset{\sim}{y} = \underset{\sim}{y}(t,\varepsilon)$ as $\varepsilon \to 0^+$ satisfying in $[0,1]$

$$||\underset{\sim}{y}(t,\varepsilon)|| \leq ||\underset{\sim}{A}||\exp[-mt/\varepsilon] + ||B||\ \exp[-m(1-t)/\sqrt{\varepsilon}],$$

for $m^2 = \min\{1-3A_i^2,\ 1-3B_i^2\}$, $i = 1,2$.

The restriction imposed on $\underset{\sim}{A}$ and $\underset{\sim}{B}$ is rather severe; we can try to relax it slightly by replacing the strong positive definiteness assumption with the weaker integral condition alluded to in Chapter VII. A short calculation shows that

$$\underset{\sim}{y}^T\ \underset{\sim}{h}(\underset{\sim}{y})/||\underset{\sim}{y}|| \geq (||\underset{\sim}{y}||^2 - ||\underset{\sim}{y}||^4)/||\underset{\sim}{y}|| = ||\underset{\sim}{y}|| - ||\underset{\sim}{y}||^3,$$

where we have used the simple inequality $y_1^4 + y_2^4 \leq (y_1^2 + y_2^2)^2$. Thus, by applying the reasoning in Example 8.3 to $||\underset{\sim}{y}||$, we conclude that, in fact, the solution $\underset{\sim}{y} = \underset{\sim}{y}(t,\varepsilon)$ found earlier actually exists and satisfies $\lim_{\varepsilon\to 0^+} \underset{\sim}{y}(t,\varepsilon) = (0\ 0)^T$ in $[\delta, 1-\delta]$ for boundary values such that $||\underset{\sim}{A}||$, $||\underset{\sim}{B}|| < \sqrt{2}$. These bounds are sharper than the bounds $||\underset{\sim}{A}||$, $||\underset{\sim}{B}|| < \sqrt{2/3}$ obtained from the more restrictive definiteness condition.

Example 8.22. Consider now the problem in $(0,1)$

$$\varepsilon y_1'' = y_1(1-y_1)(1+y_2^2) \equiv h_1(y_1,y_2), y_1(0,\varepsilon) = A_1,\ y_1(1,\varepsilon) = B_1,$$

$$\varepsilon y_2'' = y_2(1-y_2)(1+y_1^2) \equiv h_2(y_1,y_2), y_2(0,\varepsilon) = A_2,\ y_2(1,\varepsilon) = B_2,$$

in order to illustrate the componentwise theory. The function $\underset{\sim}{u} \equiv \underset{\sim}{0}$ is clearly a solution of the reduced system which satisfies, in addition, $h_1(0,y_2) \equiv 0$ for all y_2 and $h_2(y_1,0) \equiv 0$ for all y_1. Moreover, it is stable in the sense of Definition 7.2 since

$$h_{1,y_1}(0,y_2) = 1 + y_2^2 \geq 1 \quad \text{and} \quad h_{2,y_2}(y_1,0) = 1 + y_1^2 \geq 1.$$

Finally we observe that

$$h_{1,y_1}(y_1,y_2) = (1-2y_1)(1+y_2^2) > 0$$

and

$$h_{2,y_2}(y_1,y_2) = (1-2y_2)(1+y_1^2) > 0,$$

for all values of $y_1,y_2 < 1/2$. Therefore, for boundary values such that $A_i < 1/2$ and $B_i < 1/2$ $(i = 1,2)$, Theorem 7.2 states that the problem has a solution $\underset{\sim}{y} = \underset{\sim}{y}(t,\varepsilon)$ as $\varepsilon \to 0^+$ satisfying

$$\lim_{\varepsilon\to 0^+} \underset{\sim}{y}(t,\varepsilon) = (0\ 0)^T \quad \text{in} \quad [\delta,1-\delta]. \tag{8.27}$$

We note that this bound on the boundary values can be improved by using the less restrictive integral conditions that

$$\int_0^{\eta_1} h_1(s,A_2 \text{ or } B_2)ds > 0 \quad \text{and} \quad \int_0^{\eta_2} h_2(A_1 \text{ or } B_1,s)ds > 0,$$

for all values of η_i between 0 and A_i or B_i, $\eta_i \neq 0$ $(i = 1,2)$; cf. [69]. For boundary values such that $A_i,B_i < 3/2$, $i = 1,2$, we deduce therefore the existence of a solution satisfying the limiting relation (8.27).

Application 8.7. The scalar theory of Application 8.2 concerned itself, of course, with the behavior of a single reactant undergoing an isothermal catalytic reaction. Suppose however that we have a system of N reactants, each component of which undergoes such a reaction, influenced by and influencing the other $N-1$ components. Then by arguing as in Application 8.2 (cf. also [2; Chapter 5]), we see that the steady-state behavior of the concentrations can be governed by a boundary value problem of the form

$$\varepsilon \underset{\sim}{y}'' = \underset{\sim}{h}(\underset{\sim}{y}), \quad 0 < x < 1,$$

$$P\underset{\sim}{y}(0,\varepsilon) - \underset{\sim}{y}'(0,\varepsilon) = \underset{\sim}{A}, \quad Q\underset{\sim}{y}(1,\varepsilon) + \underset{\sim}{y}'(1,\varepsilon) = \underset{\sim}{B},$$

Here $\underset{\sim}{y} = (y_1 \dots y_N)^T$ is the vector of normalized concentrations, $\underset{\sim}{h}$ is an N-vector-valued function of $\underset{\sim}{y}$ which represents the nonlinear kinetics, and x is the normalized distance. P, Q are positive semidefinite $N \times N$ matrices which contain the various transfer coefficients between the bulk flow and the solid phase, and ε^2 is the reciprocal of the Thiele modulus, assumed to be the same for each reaction.

Suppose now that $\underset{\sim}{h}(\underset{\sim}{0}) = \underset{\sim}{0}$ and that the Jacobian matrix J of $\underset{\sim}{h}$ evaluated along $\underset{\sim}{0}$ is positive definite, in the sense that there is a positive constant m_1 for which

$$\underset{\sim}{y}^T J \underset{\sim}{y} \geq m_1^2 \, ||\underset{\sim}{y}||^2, \tag{8.28}$$

for all $\underset{\sim}{y}$ in $\mathbb{R}^N$. It follows from the Mean Value Theorem that

$$\underset{\sim}{y}^T \underset{\sim}{h}(\underset{\sim}{y})/||\underset{\sim}{y}|| \geq m_1^2 \, ||\underset{\sim}{y}||,$$

for $||\underset{\sim}{y}||$ sufficiently small. Consequently Theorem 7.3 tells us that the problem has a solution $\underset{\sim}{y} = \underset{\sim}{y}(x,\varepsilon)$ as $\varepsilon \to 0^+$ satisfying in $[0,1]$

$$||\underset{\sim}{y}(x,\varepsilon)|| \leq (\sqrt{\varepsilon}/m)||\underset{\sim}{A}||\exp[-mx/\sqrt{\varepsilon}] + (\sqrt{\varepsilon}/m)||\underset{\sim}{B}|| \exp[-m(1-x)/\sqrt{\varepsilon}]$$

for $0 < m < m_1$.

As an illustration, consider the problem in $(0,1)$

$$\varepsilon y_1'' = y_1(1-y_2) - (k-\lambda)y_2 \equiv h_1(y_1,y_2),$$

$$py_1(0,\varepsilon) - y_1'(0,\varepsilon) = p, \quad py_1(1,\varepsilon) + y_1'(1,\varepsilon) = p,$$

$$\varepsilon y_2'' = -y_1(1-y_2) + ky_2 \equiv h_2(y_1,y_2), \quad -y_2'(0,\varepsilon) = 0, \quad y_2'(1,\varepsilon) = 0,$$

where k, λ and p are positive constants with $k > \lambda$. It is taken from Aris's discussion [2; Chapter 5] of the pseudo-steady-state hypothesis in an enzyme reaction. Clearly $\underset{\sim}{u} = \underset{\sim}{0}$ is the only solution of the reduced equation $\underset{\sim}{h} = (h_1 \; h_2)^T = (0 \; 0)^T$, and the corresponding Jacobian matrix J of $\underset{\sim}{h}$ evaluated along $\underset{\sim}{0}$ is

$$\begin{pmatrix} 1 & -(k-\lambda) \\ -1 & k \end{pmatrix}.$$

Since k is positive, a necessary and sufficient condition for J to be positive definite is that

$$(k-\lambda+1)^2 < 4k,$$

in which case inequality (8.28) obtains with $m_1^2 = \{k+1 - [(k-1)^2 + (1-\lambda+1)^2]^{1/2}\}/2$. (Here we have used the results [26; Chapter 8] that a real symmetric 2×2 matrix (a_{ij}) is positive definite if and only if $a_{11} > 0$ and $a_{11}a_{22} - a_{12}^2 > 0$, and that $\underset{\sim}{y}^T A\underset{\sim}{y} = \underset{\sim}{y}^T A^*\underset{\sim}{y}$ for $A^* = (A + A^T)/2$.) Thus, for such values of k and λ the problem has a solution $\underset{\sim}{y} = \underset{\sim}{y}(x,\varepsilon)$ as $\varepsilon \to 0^+$ satisfying in $[0,1]$

$$||\underset{\sim}{y}(x,\varepsilon)|| \le (\sqrt{\varepsilon}p)/m\{\exp[-mx/\sqrt{\varepsilon}] + \exp[-m(1-x)/\sqrt{\varepsilon}]\},$$

for $0 < m < m_1$.

<u>Example 8.23</u>. Consider finally the symmetric problem in $(0,1)$

$$\varepsilon y_1'' = y_1-y_1^2/2-y_1y_2 \equiv h_1(y_1,y_2), y_1(0,\varepsilon) = A_1,\ y_1(1,\varepsilon) = B_1,$$

$$\varepsilon y_2'' = y_2-y_2^2/2-y_1y_2 \equiv h_2(y_1,y_2), y_2(0,\varepsilon) = A_2, y_2(1,\varepsilon) = B_2;$$

such systems arise in the study of travelling wave solutions to nonlinear partial differential equations governing the dynamics of stratified fluids (cf. [4], [82]). The reduced solution $\underset{\sim}{u} = \underset{\sim}{0}$ satisfies $h_1(0,y_2) \equiv 0$, $h_2(y_1,0) \equiv 0$ for all y_1,y_2, and $h_{1,y_1}(0,y_2) > 0$, $h_{2,y_2}(y_1,0) > 0$ for all $y_1,y_2 < 1$. Invoking the integral conditions for boundary layer behavior, we see that

$$\int_0^{A_1} h_1(s,A_2)ds > 0 \quad \text{for } A_1 < 3(1-A_2),$$

while

$$\int_0^{A_2} h_2(A_1,s)ds > 0 \quad \text{for } A_2 < 3(1-A_1),$$

and similarly for B_1,B_2. If $A_1,A_2,B_1,B_2 < 3/4$ then all of the boundary layer inequalities obtain, and so (cf. [68], [69]) the problem has a solution as $\varepsilon \to 0^+$ satisfying

$$\lim_{\varepsilon\to0^+} y(t,\varepsilon) = (0\ 0)^T \quad \text{in } [\delta,1-\delta].$$

Recalling Example 8.3 we suspect that the system also has solutions with spikes. This follows because in a subinterval of $(0,1)$ where $y_1(y_2)$ is asymptotically zero, the y_1-equation (y_2-equation) reduces to the uncoupled equation $\varepsilon y'' = y-y^2/2$, which has solutions like those of Example 8.3. More precisely, O'Malley's result [76] states that for each integer $n \geq 2$ the scalar problem

$$\begin{aligned} &\varepsilon y'' = y - y^2/2, \quad 0 < t < 1, \\ &y(0,\varepsilon) = A, \quad y(1,\varepsilon) = B, \quad A,B < 3/4, \end{aligned} \tag{*}$$

has four solutions $y = y(t,\varepsilon)$ as $\varepsilon \to 0^+$ satisfying

$$\lim_{\varepsilon\to 0^+} y(t,\varepsilon) = 0 \quad \text{in} \quad [\delta, 1-\delta],$$

except that

$$\lim_{\varepsilon\to 0^+} y(t_\ell,\varepsilon) = 3 \quad \text{for} \quad t_\ell = \ell/n, \quad \ell = 1,\ldots,n-1.$$

It follows from the theory of Chapter VII, for example by taking solutions of (*) for $n = 2$ and $n = 3$, that the original vector problem has the eight solutions pictured below, together with the eight solutions obtained from these by interchanging y_1 and y_2. Using the solutions of (*) for $n = 2,3,\ldots,p$, p a prime, we can show the existence of solutions as $\varepsilon \to 0^+$ having spikes of height 3 at the t-values 1/2, 1/3, 2/3, 1/5, 2/5, 3/5, 4/5,...,1/p, 2/p,...,(p-1)/p, provided that A_1, B_1, A_2, $B_2 < 3/4$.

§8.6. Examples of Quasilinear Systems and An Application

Example 8.24. We turn first to an illustration of a norm-bound result for a problem with a solution exhibiting boundary layer behavior, namely

$$\begin{aligned} &\varepsilon \underset{\sim}{y}'' = F(\underset{\sim}{y})\underset{\sim}{y}' + \underset{\sim}{g}(\underset{\sim}{y}), \quad 0 < t < 1, \\ &\underset{\sim}{y}(0,\varepsilon) = \underset{\sim}{A}, \quad \underset{\sim}{y}(1,\varepsilon) = \underset{\sim}{B}. \end{aligned}$$

Here $\underset{\sim}{y} = (y_1\ y_2)^T$, $F = -\begin{pmatrix} y_1 & 1 \\ 1 & y_2 \end{pmatrix}$, $\underset{\sim}{g} = (y_1+1\ \ y_2+1)^T$, and so this problem can be regarded as a generalization of the Lagerstrom-Cole model problem (cf. Example 8.10) to two dimensions. Let us examine under what conditions there is a solution with a boundary layer at $t = 0$. We begin by assuming that the reduced problem $F(\underset{\sim}{u}_R)\underset{\sim}{u}_R' + \underset{\sim}{g}(\underset{\sim}{u}_R) = \underset{\sim}{0}$, $\underset{\sim}{u}_R(1) = \underset{\sim}{B}$, has a stable solution in [0,1], that is, the matrix $F(\underset{\sim}{u}_R(t))$ must be negative definite. We also require that there is boundary layer stability at $t = 0$, in that the matrix $F(\underset{\sim}{u}_R(0) + \underset{\sim}{\xi})$ must be negative definite for all 2-vectors $\underset{\sim}{\xi}$ satisfying $0 < ||\underset{\sim}{\xi}|| \leq ||\underset{\sim}{A}-\underset{\sim}{u}_R(0)||$.

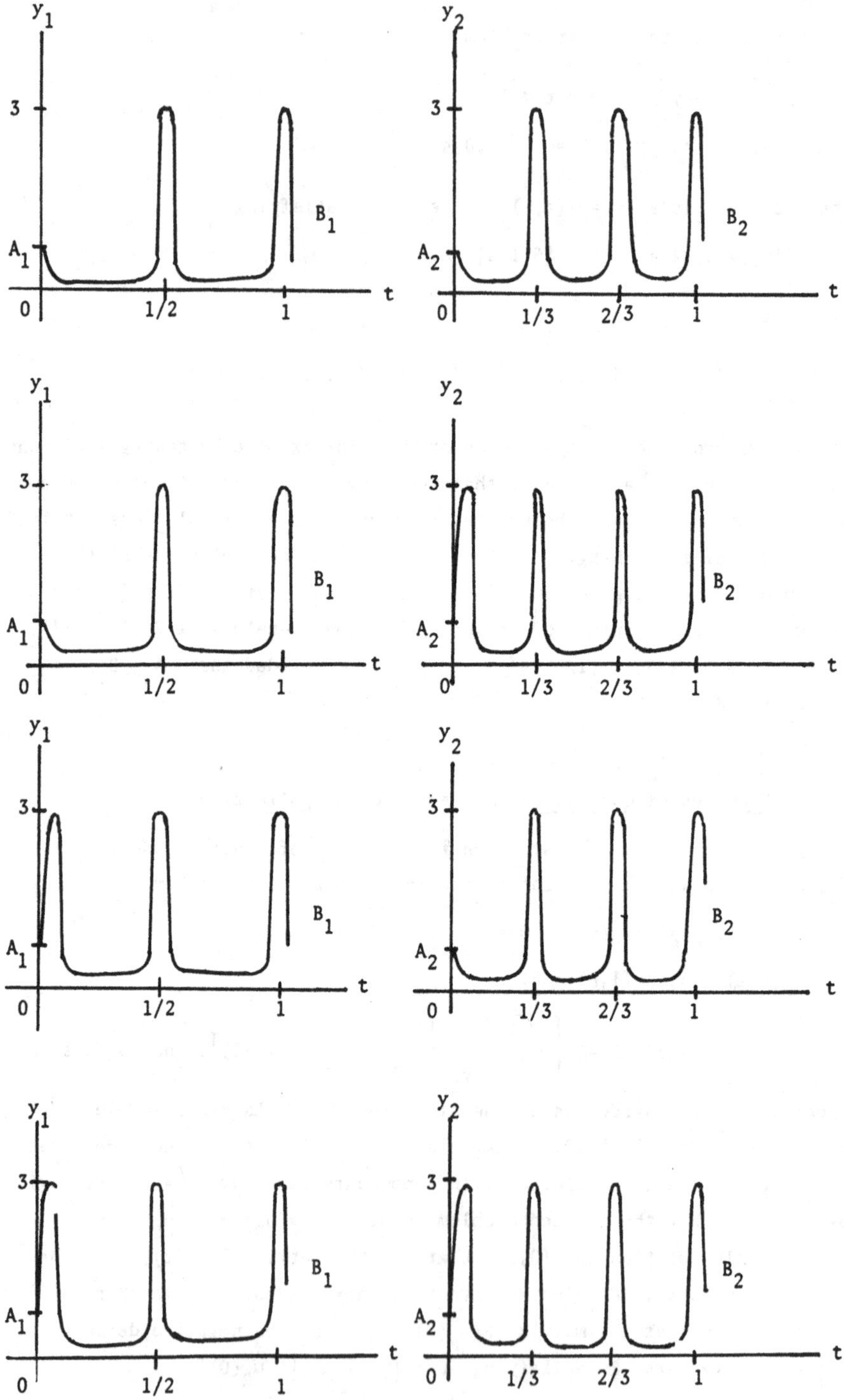

Figure 8.6
Eight Solutions of Example 8.23

The reduced problem has the solution $\underset{\sim}{u}_R(t) = (t+C\ \ t+D)^T$, for $(C\ D) = (B_1-1\ \ B_2-1)$, and we see that $\underset{\sim}{u}_R$ is stable in $[0,1]$ provided

$$\begin{bmatrix} t+C & 1 \\ 1 & t+D \end{bmatrix} > 0, \text{ that is, provided } C+D > 0 \text{ and } CD > 1, \text{ or}$$

$B_1 B_2 > B_1 + B_2 > 2$. Finally, boundary layer stability requires that

$$\begin{bmatrix} \xi_1 + C & 1 \\ 1 & \xi_2 + D \end{bmatrix} > 0, \text{ that is,}$$

$\xi_1^3 + C\xi_1^2 + 2\xi_1\xi_2 + \xi_2^3 + D\xi_2^2 > 0$, for all $(\xi_1\ \xi_2)^T$ satisfying $0 < ||\underset{\sim}{\xi}|| <$ $||\underset{\sim}{A}-\underset{\sim}{u}_R(0)|| = \sqrt{(A_1-B_1+1)^2 + (A_2-B_2+1)^2}$. Thus the initial values $\underset{\sim}{y}(0,\varepsilon) = \underset{\sim}{A}$ are restricted to a disk about $(C,D) = \underset{\sim}{u}_R(0)$ whose radius is less than the least norm $||\underset{\sim}{\xi}||$ of the nontrivial zeros of the cubic polynomial. Setting $\xi_2 = t\xi_1$ such a $\underset{\sim}{\xi}$ will satisfy $(1+t^3)\xi_1 = -(C + 2t + Dt^2)$, and we minimize $d(t) = ||\underset{\sim}{\xi}|| = \sqrt{1 + t^2}\ |\xi_1|$. The solution of this calculus problem then determines an upper bound for $||\underset{\sim}{A} - \underset{\sim}{u}_R(0)||$.

For $C = D = 2$, that is, $\underset{\sim}{B} = (3\ 3)^T$, we obtain the minimum value 2 for $d(t)$, corresponding to $t_{min} = 0$. Thus Theorem 7.4 tells us that if $\underset{\sim}{A}$ lies in the disk of radius 2 about $(2\ 2)^T$, then the problem has a solution $\underset{\sim}{y} = \underset{\sim}{y}(t,\varepsilon)$ as $\varepsilon \to 0^+$ satisfying

$$\lim_{\varepsilon\to 0^+} \underset{\sim}{y}(t,\varepsilon) = \underset{\sim}{u}_R(t) \quad \text{in } [\delta,1]. \tag{8.29}$$

This is a rather severe restriction on the size of the boundary layer jump $||\underset{\sim}{A} - \underset{\sim}{u}_R(0)||$. It can be improved by replacing the definiteness condition on $F(\underset{\sim}{u}_R(0) + \underset{\sim}{\xi})$ with the less demanding integral condition that

$$\underset{\sim}{\xi}^T \int_{\underset{\sim}{0}}^{\underset{\sim}{\xi}} F(\underset{\sim}{u}_R(0) + \underset{\sim}{\xi})d\underset{\sim}{\xi} < 0,$$

for all 2-vectors $\underset{\sim}{\xi}$ such that $0 < ||\underset{\sim}{\xi}|| \leq ||\underset{\sim}{A} - \underset{\sim}{u}_R(0)||$; cf. [47]. In our case this integral condition is equivalent to the requirement that

$$\xi_1^3 + 2C\xi_1^2 + 4\xi_1\xi_2 + \xi_2^3 + 2D\xi_2^3 > 0,$$

for all such $\underset{\sim}{\xi}$. Proceeding as before, we can show that for $C = D = 2$ the minimum value of $d(t)$ is 3.39, corresponding to $t_{min} = -0.291$. Thus if $\underset{\sim}{A}$ lies in the disk of radius 3.39 about $(2\ 2)^T$, then the limiting relation (8.29) obtains again. Even though this is an improvement over the previous result, the estimate on $||\underset{\sim}{A} - \underset{\sim}{u}_R(0)||$ is nowhere near optimal. We expect that boundary layer stability need only hold for $\underset{\sim}{\xi} + \underset{\sim}{u}_R(0)$ on the actual solution path joining $\underset{\sim}{A}$ and $\underset{\sim}{u}_R(0)$.

Example 8.25. We consider now an example which illustrates the componentwise boundary layer results, as well as the differences between this theory and the norm-bound theory, namely

$$\varepsilon y_1'' = (1-y_1)y_1' + y_1 g_1(t,y_2), y_1(0,\varepsilon) = 0, y_1(1,\varepsilon) = B_1,$$

$$\varepsilon y_2'' = (1-y_2)y_2' + y_2 g_2(t,y_1), y_2(0,\varepsilon) = 0, y_2(1,\varepsilon) = B_2,$$

for t in $(0,1)$. The lefthand reduced problem clearly has the solution $u_1 = u_2 \equiv 0$, which is stable because $f_1(0) = f_2(0) > 0$ for $f_i(y_i) = 1-y_i$, $i = 1,2$. Since $f_i(y_i) > 0$ whenever $y_i < 1$, we conclude from Theorem 7.5 that the problem has a solution $\underset{\sim}{y} = \underset{\sim}{y}(t,\varepsilon)$ as $\varepsilon \to 0^+$, for all $B_1, B_2 < 1$, satisfying in $[0,1]$

$$y_i(t,\varepsilon) = B_i \exp[-k(1-t)/\varepsilon],$$

where $0 < k < \min\{1-B_1, 1-B_2\}$.

The restriction on $\underset{\sim}{B}$ can be relaxed by using appropriate integral conditions, as was done in the scalar theory. We require that

$$(u_i(1) - B_i)\int_{u_i(1)}^{\xi} f_i(\underset{\sim}{B}_{si})ds < 0, \tag{8.30}$$

for $B_i \le \xi < u_i(1)$ if $B_i < u_i(1)$ or for $u_i(1) < \xi \le B_i$ if $u_i(1) < B_i$. For both components we see that

$$\int_{u_i(1)}^{\xi} f_i(\underset{\sim}{B}_{si})ds = \int_0^{\xi}(1-s)ds = \xi - \xi^2/2 > 0$$

whenever $0 < \xi < 2$, and therefore (8.30) obtains for $-\infty < B_i < 2$. A theorem of O'Donnell [68] tells us that the solution satisfies

$$\lim_{\varepsilon\to 0^+} \underset{\sim}{y}(t,\varepsilon) = \underset{\sim}{0} \quad \text{in} \quad [0,1-\delta],$$

for all $B_1, B_2 < 2$.

Let us now compare this result with that obtained from the norm-bound theory. This theory requires that $F(\underset{\sim}{0}) > 0$, for $F(\underset{\sim}{y}) = \mathrm{diag}\{1-y_1, 1-y_2\}$, and that the inner product

$$\underset{\sim}{y}^T \int_0^{\underset{\sim}{y}} F(\underset{\sim}{s}) d\underset{\sim}{s} > 0,$$

for all $\underset{\sim}{y}$ on paths between $\underset{\sim}{0}$ and $\underset{\sim}{B}$ satisfying $0 < ||\underset{\sim}{y}|| \le ||\underset{\sim}{B}||$. The stability condition is certainly satisfied since $F(\underset{\sim}{0})$ is the identity matrix, while the boundary layer stability condition is

$$\underset{\sim}{y}^T \int_0^{\underset{\sim}{y}} \text{diag}\{1-s_1, 1-s_2\} (ds_1 \ ds_2)^T > 0,$$

that is, $y_1^2(1-y_1/2) + y_2^2(1-y_2/2) > 0$, for all $\underset{\sim}{y} = (y_1 \ y_2)^T$, $0 < ||\underset{\sim}{y}|| \le ||\underset{\sim}{B}||$. Thus, we must require that $||\underset{\sim}{B}|| < 2$, even though $y_1^2(1-y_1/2) + y_2^2(1-y_2/2)$ is positive whenever $y_1, y_2 < 2$.

Application 8.8. As an application of the quasilinear theory consider the following system analog of Application 8.1. A reactant fluid flows through a tubular reactor at a constant average speed U, and we assume that there is axial dispersion caused by turbulent mixing. If N species react isothermally, then the N-vector $\underset{\sim}{y}$ of steady-state concentrations satisfies a system of the form (cf. [80; Chapter 4])

$$\varepsilon \underset{\sim}{y}'' = U \underset{\sim}{y}' + \underset{\sim}{g}(\underset{\sim}{y}), \quad 0 < x < L,$$

$$\underset{\sim}{y}(0,\varepsilon) = \underset{\sim}{y}_0, \quad \underset{\sim}{y}(L,\varepsilon) = \underset{\sim}{y}_L.$$

Here ε is the reciprocal of the Peclet number, $\underset{\sim}{g}$ is an N-vector-valued function which contains the kinetics of the reaction, L is the length of the tubular reactor, and $\underset{\sim}{y}_0$ and $\underset{\sim}{y}_L$ are the prescribed values of the reactant concentrations at the inlet and outlet of the tube, respectively. Since U is positive we look for a solution $\underset{\sim}{u} = \underset{\sim}{u}(x)$ of the lefthand reduced problem

$$U \underset{\sim}{u}' + \underset{\sim}{g}(\underset{\sim}{u}) = \underset{\sim}{0}, \quad \underset{\sim}{u}(0) = \underset{\sim}{y}_0.$$

Let us assume that $\underset{\sim}{u}$ exists in $[0,L]$; then $\underset{\sim}{u}$ is clearly stable, in the sense of both Definition 7.4 and Definition 7.5. We also have boundary layer stability at $x = L$ in both senses. Thus the norm-bound theory tells us that the original problem has a solution $\underset{\sim}{y} = \underset{\sim}{y}(x,\varepsilon)$ as $\varepsilon \to 0^+$ satisfying in $[0,L]$

$$||\underset{\sim}{y}(x,\varepsilon) - \underset{\sim}{u}(x)|| \le ||\underset{\sim}{y}_L - \underset{\sim}{u}(1)|| \ \exp[-U(L-x)/\varepsilon] + 0(\varepsilon), \quad (8.31)$$

that is, there is a boundary layer at $x = L$. Moreover, the componentwise theory provides the sharper estimate, for x in $[0,L]$ and $1 \le i \le N$,

$$|y_i(x,\varepsilon) - u_i(x)| \le |b_i - u_i(L)| \exp[-U(L-x)/\varepsilon] + O(\varepsilon), \tag{8.32}$$

where b_i is the i-th component of $\underset{\sim}{y}_L$. If the Dirichlet condition at $x = L$ is replaced with a zero-flux condition, $\underset{\sim}{y}'(L,\varepsilon) = \underset{\sim}{0}$, then we expect that the solution is uniformly close to $\underset{\sim}{u}(x)$ in $[0,L]$; cf. Application 8.4.

It is also possible to consider the more general situation in which each species y_i flows at a constant average speed $U_i > 0$, with $U_i \neq U_j$, in general, for $i \neq j$. The differential equation for $\underset{\sim}{y}$ becomes

$$\varepsilon \underset{\sim}{y}'' = \mathrm{diag}\{U_1,\ldots,U_N\}\underset{\sim}{y}' + \underset{\sim}{g}(\underset{\sim}{y}),$$

and clearly the above theories apply mutatis mutandis. In particular, the estimate (8.31) obtains with U replaced by $\min_{1\le i\le N} U_i$, while each of the estimates in (8.32) obtains with U replaced by U_i.

References

[1] R. C. Ackerberg and R. E. O'Malley, Jr., Boundary Layer Problems Exhibiting Resonance, Studies in Appl. Math. 49 (1970), 277-295.

[2] R. Aris, The Mathematical Theory of Diffusion and Reaction in Permeable Catalysts, Vol. I, Clarendon Press, Oxford, 1975.

[3] G. K. Batchelor, An Introduction to Fluid Dynamics, Cambridge Univ. Press, Cambridge, 1967.

[4] D. J. Benney, Long Non-Linear Waves in Fluid Flows, J. Math. and Phys. 45 (1966), 52-62.

[5] S. R. Bernfeld and V. Lakshmikantham, An Introduction to Nonlinear Boundary Value Problems, Academic Press, New York, 1974.

[6] Yu. P. Boglaev, The Two-Point Problem for a Class of Ordinary Differential Equation with a Small Parameter Coefficient of the Derivative, USSR Comp. Math. and Math. Phys. 10 (1970), 4, 191-204.

[7] N. I. Bris, On Boundary Value Problems for the Equation $\varepsilon y'' = f(x,y,y')$ for Small ε (in Russian), Dokl. Akad. Nauk SSSR 95 (1954), 429-432.

[8] G. F. Carrier, Singular Perturbation Theory and Geophysics, SIAM Rev. 12 (1970), 175-193.

[9] K. W. Chang, On Coddington and Levinson's Results for a Nonlinear Boundary Value Problem Involving a Small Parameter, Rend. Accad. Nazionale dei Lincei 54 (1973), 356-363.

[10] K. W. Chang, Approximate Solutions of Nonlinear Boundary Value Problems Involving a Small Parameter, SIAM J. Appl. Math. 30 (1976), 42-54.

[11] K. W. Chang, Diagonalization Method for a Vector Boundary Problem of Singular Perturbation Type, J. Math. Anal. Appl. 48 (1974), 652-665.

[12] K. W. Chang, Singular Perturbations of a Boundary Problem for a Vector Second Order Differential Equation, SIAM J. Appl. Math. 30 (1976), 42-54.

[13] K. W. Chang and W. A. Coppel, Singular Perturbations of Initial Value Problems over a Finite Interval, Arch. Rational Mech. Anal. 32 (1969), 268-280.

[14] E. A. Coddington and N. Levinson, A Boundary Value Problem for a Nonlinear Differential Equation with a Small Parameter, Proc. Amer. Math. Soc. 3 (1952), 73-81.

[15] D. S. Cohen, Singular Perturbation of Nonlinear Two-Point Boundary-Value Problems, J. Math. Anal. Appl. 43 (1973), 151-160.

[16] L. P. Cook and W. Eckhaus, Resonance in a Boundary Value Problem of Singular Perturbation Type, Studies in Appl. Math. 52 (1973), 129-139.

[17] R. Courant and K. O. Friedrichs, Supersonic Flow and Shock Waves, Springer-Verlag, New York, 1976.

[18] L. Crocco, A Suggestion for the Numerical Solution of the Steady Navier-Stokes Equations, AIAA J. 3 (1965), 1824-1832.

[19] M. A. Deshpande, Perturbations and Applications, Doctoral Diss., Marathwada Univ., Aurangabad, India, 1981.

[20] F. W. Dorr, S. V. Parter and L. F. Shampine, Applications of the Maximum Principle to Singular Perturbation Problems, SIAM Rev. 15 (1973), 43-88.

[21] W. Eckhaus, Asymptotic Analysis of Singular Perturbations, North-Holland, Amsterdam, 1979.

[22] A. Erdélyi, On a Nonlinear Boundary Value Problem Involving a Small Parameter, J. Austral. Math. Soc. 2 (1962), 425-439.

[23] A. Erdélyi, Singular Perturbations of Boundary Value Problems Involving Ordinary Differential Equations, SIAM J. Appl. Math. 11 (1963), 105-116.

[24] P. C. Fife, Semilinear Elliptic Boundary Value Problems with Small Parameters, Arch. Rational Mech. Anal. 52 (1973), 205-232.

[25] J. E. Flaherty and R. E. O'Malley, Jr., The Numerical Solution of Boundary Value Problems for Stiff Differential Equations, Math. Comp. 31 (1977), 66-93.

[26] F. R. Gantmacher, The Theory of Matrices, Vol. I, Chelsea, New York, 1960.

[27] S. Haber and N. Levinson, A Boundary Value Problem for a Singularly Perturbed Differential Equation, Proc. Amer. Math. Soc. 6 (1955), 866-872.

[28] P. Habets, Singular Perturbations of Nonlinear Boundary Value Problems, Lecture Notes, Catholic Univ. of Louvain, 1974.

[29] P. Habets, Double Degeneracy in Singular Perturbation Problems (in French), Annales Soc. Scientifique de Bruxelles 89 (1975), 11-15.

[30] P. Habets, Singular Perturbations of a Vector Boundary Value Problem, in Lectures Notes in Math., no. 415, Springer-Verlag, New York, 1974, pp. 149-154.

[31] P. Habets and M. Laloy, Etude de Problemes aux Limites par la Methode des Sur-et Sous-Solutions, Lecture Notes, Catholic Univ. of Louvain, 1974.

[32] W. A. Harris, Jr., Singular Perturbations of a Boundary Value Problem for a System of Differential Equations, Duke Math. J. 29 (1962), 429-445.

[33] A. van Harten, Singular Perturbation Problems for Non-Linear Elliptic Second Order Equations, in North-Holland Math. Studies, Vol. 13, 1974, pp. 181-195.

[34] P. Hartman, On Boundary Value Problems for Systems of Ordinary, Nonlinear, Second Order Differential Equations, Trans. Amer. Math. Soc. 96 (1960), 493-509.

[35] P. Hartman, Ordinary Differential Equations, Hartman, Baltimore, 1973.

[36] J. W. Heidel, A Second-Order Nonlinear Boundary Value Problem, J. Math. Anal. Appl. 48 (1974), 493-503.

[37] F. Hoppensteadt, Properties of Solutions of Ordinary Differential Equations with a Small Parameter, Comm. Pure Appl. Math. 24 (1970), 807-840.

[38] F. A. Howes, Singularly Perturbed Nonlinear Boundary-Value Problems Whose Reduced Equations Have Singular Points, Studies in Appl. Math. 57 (1977), 135-180.

[39] F. A. Howes, Boundary-Interior Layer Interactions in Nonlinear Singular Perturbation Theory, Memoirs Amer. Math. Soc., no. 203, 1978.

[40] F. A. Howes, An Asymptotic Theory for a Class of Nonlinear Robin Problems, Part I [Part II], J. Differential Equations 30 (1978), 192-234 [Trans. Amer. Math. Soc. 260 (1980), 527-552].

[41] F. A. Howes, Singularly Perturbed Semilinear Systems, Studies in Appl. Math. 61 (1979), 185-209.

[42] F. A. Howes, Singularly Perturbed Superquadratic Boundary Value Problems, Nonlinear Analysis 3 (1979), 175-192.

[43] F. A. Howes, Some Old and New Results on Singularly Perturbed Boundary Value Problems, in Singular Perturbations and Asymptotics, ed. by R. E. Meyer and S. V. Parter, Academic Press, New York, 1980, pp. 41-85.

[44] F. A. Howes, Some Singularly Perturbed Superquadratic Boundary Value Problems Whose Solutions Exhibit Boundary and Shock Layer Behavior, Nonlinear Analysis 4 (1980), 683-698.

[45] F. A. Howes, Differential Inequality Techniques and Singular Perturbations, Rocky Mtn. J. Math. 12 (1982), 767-777.

[46] F. A. Howes, Singularly Perturbed Semilinear Robin Problems, Studies in Appl. Math. 67 (1982), 125-139.

[47] F. A. Howes and R. E. O'Malley, Jr., Singular Perturbations of Second-Order Semilinear Systems, in Lecture Notes in Math., no. 827, Springer-Verlag, New York, 1980, pp. 131-150.

[48] E. L. Ince, Ordinary Differential Equations, Dover, New York, 1956.

[49] L. K. Jackson, Subfunctions and Second-Order Ordinary Differential Inequalities, Adv. in Math. 2 (1968), 308-363.

[50] W. E. Johnson and L. M. Perko, Interior and Exterior Boundary Value Problems from the Theory of the Capillary Tube, Arch. Rational Mech. Anal. 29 (1968), 125-143.

[51] H. B. Keller, Existence Theory for Multiple Solutions of a Singular Perturbation Problem, SIAM J. Math. Anal. 3 (1972), 86-92.

[52] W. G. Kelley, Second Order Systems with Nonlinear Boundary Conditions, Proc. Amer. Math. Soc. 62 (1977), 287-292.

[53] W. G. Kelley, A Geometric Method of Studying Two Point Boundary Value Problems for Second Order Systems, Rocky Mtn. J. Math. 7 (1977), 251-263.

[54] W. G. Kelley, A Nonlinear Singular Perturbation Problem for Second Order Systems, SIAM J. Math. Anal. 10 (1979), 32-37.

[55] J. Kevorkian and J. D. Cole, Perturbation Methods in Applied Mathematics, Springer-Verlag, New York, 1981.

[56] H. O. Kreiss and S. V. Parter, Remarks on Singular Perturbations with Turning Points, SIAM J. Math. Anal. 5 (1974), 230-251.

[57] A. Lasota and J. A. Yorke, Existence of Solutions of Two Point Boundary Value Problems for Nonlinear Systems, J. Differential Equations 11 (1972), 509-518.

[58] T. C. Lee, Boundary Value Problems for Second Order Ordinary Differential Equations and Applications to Singular Perturbation Problems on $[a,b] \subset (-\infty,\infty]$, Rocky Mtn. J. Math. 6 (1976), 761-765.

[59] J. J. Levin, Singular Perturbations of Nonlinear Systems of Differential Equations Related to Conditional Stability, Duke Math. J. 23 (1956), 609-620.

[60] J. J. Levin, The Asymptotic Behavior of the Stable Initial Manifolds of a System of Nonlinear Differential Equations, Trans. Amer. Math. Soc. 85 (1957), 357-368.

[61] J. J. Levin and N. Levinson, Singular Perturbations of Nonlinear Systems and an Associated Boundary Layer Equation, J. Rational Mech. Anal. 3 (1954), 247-270.

[62] J. W. Macki, Singular Perturbations of a Boundary Value Problem for a System of Ordinary Differential Equations, Arch. Rational Mech. Anal. 24 (1967), 219-232.

[63] B. J. Matkowsky, On Boundary Layer Problems Exhibiting Resonance, SIAM Rev. 17 (1975), 82-100; Errata, ibid. 18 (1976), 112.

[64] R. von Mises, Die Grenzschichte in der Theorie der gewöhnlichen Differentialgleichungen, Acta Univ. Szeged., Sect. Sci. Math. 12 (1950), 29-34.

[65] A. D. Myshkis and G. V. Scherbina, A Limit Boundary Value Problem Not Satisfying S. N. Bernstein's Condition with Applications in the Theory of Capillary Phenomena, Differential Equations 12 (1976), 698-704.

[66] M. Nagumo, Über die Differentialgleichung $y'' = f(x,y,y')$, Proc. Phys. Math. Soc. Japan 19 (1937), 861-866.

[67] M. Nagumo, Über das Verhalten der Integrale von $\lambda y'' + f(x,y,y',\lambda) = 0$ für $\lambda \to 0$, ibid. 21 (1939), 529-534.

[68] M. A. O'Donnell, Boundary and Interior Layer Behavior in Singularly Perturbed Systems of Boundary Value Problems, Doctoral Diss., U.C. Davis, 1983.

[69] M. A. O'Donnell, Boundary and Corner Layer Behavior in Singularly Perturbed Semilinear Systems of Boundary Value Problems, SIAM J. Math. Anal. 15 (1984), 317-332.

[70] O. A. Oleinik and A. N. Zizina, On Boundary Value Problems for the Equation $\varepsilon y'' = F(x,y,y')$ for Small ε (in Russian), Mat. Sbornik 31 (1952), 707-717.

[71] F. W. J. Olver, Uniform Asymptotic Expansions and Singular Perturbations, in SIAM-AMS Proceedings, vol. 10, 1976, pp. 105-117.

[72] R. E. O'Malley, Jr., A Boundary Value Problem for Certain Nonlinear Second Order Differential Equations with a Small Parameter, Arch. Rational Mech. Anal. 29 (1968), 66-74.

[73] R. E. O'Malley, Jr., Topics in Singular Perturbations, Adv. in Math. 2 (1968), 365-470.

[74] R. E. O'Malley, Jr., On Singular Perturbation Problems with Interior Nonuniformities, J. Math. Mech. 19 (1970), 1103-1112.

[75] R. E. O'Malley, Jr., Introduction to Singular Perturbations, Academic Press, New York, 1974.

[76] R. E. O'Malley, Jr., Phase-Plane Solutions to Some Singular Perturbation Problems, J. Math. Anal. Appl. 54 (1976), 449-466.

[77] C. E. Pearson, On Non-Linear Ordinary Differential Equations of Boundary Layer Type, J. Math. and Phys. 47 (1968), 351-358.

[78] L. M. Perko, Boundary Layer Analysis of the Wide Capillary Tube, Arch. Rational Mech. Anal. 45 (1972), 120-133.

[79] L. M. Perko, A Class of Superquadratic Boundary Value Problems of Capillary Type, ibid., 74 (1980), 355-377.

[80] E. E. Petersen, Chemical Reaction Analysis, Prentice-Hall, Englewood Cliffs, New Jersey, 1965.

[81] M. H. Protter and H. F. Weinberger, Maximum Principles in Differential Equations, Prentice-Hall, Englewood Cliffs, New Jersey, 1967.

[82] L. G. Redekopp and P. Weidman, Solitary Rossby Waves in Zonal Shear Flows and Their Interactions, J. Atmospheric Sci. 35 (1978), 790-804.

[83] J. Schröder, Operator Inequalities, Academic Press, New York, 1980.

[84] J. W. Searl, Extensions of a Theorem of Erdélyi, Arch. Rational Mech. Anal. 50 (1973), 127-138.

[85] Y. Tschen, Über das Verhalten der Lösungen einer Folge von Differentialgleichungen, welche im Limes ausarten, Compositio Math. 2 (1935), 378-401.

[86] V. A. Tupchiev, On the Existence, Uniqueness and Asymptotic Behavior of the Solution of a Boundary Value Problem for a System of Differential Equations with a Small Parameter Multiplying the Highest Derivatives, Sov. Math. Sokl. 3 (1962), 302-305.

[87] A. B. Vasil'eva, Asymptotic Behavior of Solutions of Problems for Ordinary Nonlinear Differential Equations with a Small Parameter Multiplying the Highest Derivatives, Russian Math. Surveys 18 (1963), 13-84.

[88] A. B. Vasil'eva and V. F. Butuzov, Asymptotic Expansions of Solutions of Singularly Perturbed Equations (in Russian), Nauka, Moscow, 1973.

[89] A. B. Vasil'eva and V. A. Tupchiev, Periodic Nearly-Discontinuous Solutions of Systems of Differential Equations with a Small Parameter in the Derivatives, Soc. Math. Dokl. 9 (1968), 179-183.

[90] M. I. Vishik and L. A. Liusternik, Initial Jump for Nonlinear Differential Equations Containing a Small Parameter, Sov. Math. Dokl. 1 (1960), 749-752.

[91] V. M. Volosov, Averaging in Systems of Ordinary Differential Equations, Russian Math. Surveys 17 (1962), 1-126.

[92] W. R. Wasow, Singular Perturbation of Boundary Value Problems for Nonlinear Differential Equations of the Second Order, Comm. Pure Appl. Math. 9 (1956), 93-116.

[93] W. R. Wasow, Asymptotic Expansions for Ordinary Differential Equations, Interscience, New York, 1965.

[94] W. R. Wasow, The Capriciousness of Singular Perturbations, Nieuw Arch. Wisk. 18 (1970), 190-210.

[95] V. W. Weekman and R. L. Gorring, Influence of Volume Change on Gas-Phase Reactions in Porous Catalysts, J. Catalysis 4 (1965), 260-270.

[96] D. Willett, On a Nonlinear Boundary Value Problem with a Small Parameter Multiplying the Highest Derivative, Arch. Rational Mech. Anal. 23 (1966), 276-287.

[97] F. A. Williams, Theory of Combustion in Laminar Flows, Ann. Rev. Fluid Mech., vol. 3, 1971, pp. 171-188.

Author Index

Subject Index

Applied Mathematical Sciences

cont. from page ii

36. Bengtsson/Ghil/Källén: **Dynamic Meterology: Data Assimilation Methods.**
37. Saperstone: **Semidynamical Systems in Infinite Dimensional Spaces.**
38. Lichtenberg/Lieberman: **Regular and Stochastic Motion.**
39. Piccinini/Stampacchia/Vidossich: **Ordinary Differential Equations in R^n.**
40. Naylor/Sell: **Linear Operator Theory in Engineering and Science.**
41. Sparrow: **The Lorenz Equations: Bifurcations, Chaos, and Strange Attractors.**
42. Guckenheimer/Holmes: **Nonlinear Oscillations, Dynamical Systems and Bifurcations of Vector Fields.**
43. Ockendon/Tayler: **Inviscid Fluid Flows.**
44. Pazy: **Semigroups of Linear Operators and Applications to Partial Differential Equations.**
45. Glashoff/Gustafson: **Linear Optimization and Approximation: An Introduction to the Theoretical Analysis and Numerical Treatment of Semi-Infinite Programs.**
46. Wilcox: **Scattering Theory for Diffraction Gratings.**
47. Hale et al.: **An Introduction to Infinite Dimensional Dynamical Systems — Geometric Theory.**
48. Murray: **Asymptotic Analysis.**
49. Ladyzhenskaya: **The Boundary-Value Problems of Mathematical Physics.**
50. Wilcox: **Sound Propagation in Stratified Fluids.**
51. Golubitsky/Schaeffer: **Bifurcation and Groups in Bifurcation Theory, Vol. I.**
52. Chipot: **Variational Inequalities and Flow in Porous Media.**
53. Majda: **Compressible Fluid Flow and Systems of Conservation Laws in Several Space Variables.**
54. Wasow: **Linear Turning Point Theory.**
55. Yosida: **Operational Calculus: A Theory of Hyperfunctions.**
56. Chang/Howes: **Nonlinear Singular Perturbation Phenomena: Theory and Applications.**